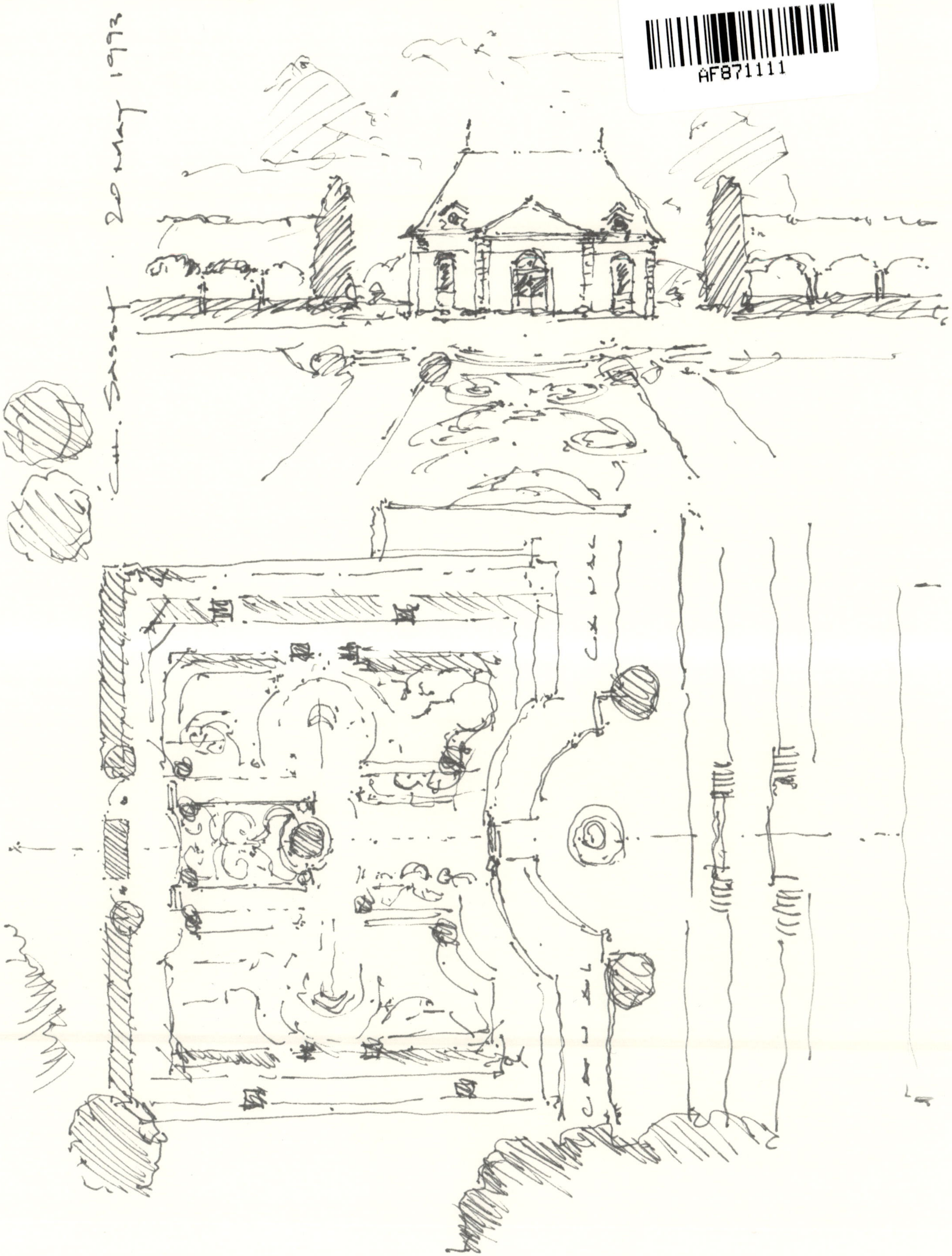

A GREAT JOY ON A RAINY AND COOL DAY IN NORMANDY

CHARLES J. STICK

AND HIS GARDENS

CHARLES J. STICK

AND HIS GARDENS

JEFF POOLE

Foreword by Sam Abell

CONTENTS

FOREWORD

In a Charles Stick garden, you are never alone. Charles is with you. He abides in the gardens he creates, invisibly but firmly guiding your experience. It's by his design that you walk where you walk and pause, sit, and stand where you do. This is immersion, not only in a garden but within the mind and spirit of its creator.

Above all, Charles shapes your seeing. Turn a corner and be drawn forward by the presence of a perfectly proportioned statue. Emerge from the garden and look back. That deeply satisfying composition leading to a distant focal point is not unintended. Charles has brought that far point, and your experience of it, into existence.

So the best way to understand and appreciate Charles is to enter one of his gardens and allow him to guide you. The second best way is to immerse yourself in this book—the first to document the life and work of Charles Stick.

Over the past twenty-five years I have come to know Charles in many—but not all—of his personas. Many because we've done much together. Not all because Charles ceaselessly reinvents himself and his place in the world. With Charles there's always more to discover—as this book proves.

Books. For several years in our town a reclusive rare book dealer ran a small, refined bookshop with highly irregular hours. One evening the shop's light was unexpectedly on. I went in.

"How's business?"

"Nonexistent. I have one customer."

"Charles J. Stick!"

I knew that because to visit Charles, either at his office or at home, is to be embraced by elements of his large and ever-growing library—substantial in size and in significance. The subject matter is broad: art and architecture, gardens and design, history and science, culture and travel. Especially travel. For Charles is a thoughtful traveler, one intent on enlarging his knowledge, especially about landscape design, from close connection with the cultures of the world.

It's through travel that I know Charles best. Together with Charlotte Tieken, we have surveyed some of the European and American gardens that Charles considers most important to the history of landscape design. On those trips I am the photographer. But I'm also the listener. And the thoughts and stories I've heard about landscape creation are in this book.

What do I, a photographer, have in common with Charles, a landscape architect? Vision. Both of us create compositions by building deeply layered structures. That building begins on the back layer. Enter a garden with Charles and watch his gaze go directly into the distance. For him that focal point is the beginning of the composition. From there the structure of the garden builds back to the viewer in successive layers. Depth is the result.

Sam Abell at work on a foggy morning at Chateau Brécy in Normandy, photographed by Charles Stick.

But inside the layered depth are a myriad of details, all to be realized with exactness. The critically shaped hedge, subtly pruned tree, sharply cut border, finely framed bed, and gracefully draped pergola are all characteristics of a Stick-made garden. To enter one of them is to inhabit a deeply considered, highly refined realm.

What is the goal? To create serious gardens—those that bring meaning to the lives of his clients. In collaboration with them Charles must manage the complex dynamics of place, plant material, and people's aspirations. I've seen him do it in many places, for the gardens of Charles Stick constitute an archipelago that extends from the Great Lakes to the mid-Atlantic and from an island in Maine to another in Florida.

But probably the preeminent Charles Stick garden is Waverley—a co-creation with Charlotte Tieken in Somerset, Virginia. It is certainly the garden that has received the fullness of Charles's vision and his most continual care.

Waverley is a garden of themed rooms, each walled by mature boxwoods, each linked by a design that culminates in a formal organic vegetable garden. There the design ends.

Or so I thought. Then, after a year's absence, I returned to Waverley. It was spring. I roamed through the flower-filled rooms and met Charles among the vegetables. He greeted me by saying, "I have something to show you."

We left the garden and encountered a formerly barren hillside. Barren no longer! Now the hillside was a continuation of the garden, thickly planted with blooming daffodils dramatically bisected by a broad grass path lined with boxwoods.

We ascended the path to the near summit of the hill. There stood an imposing column of carved limestone surmounted by a classical urn. Turning now, we gazed backward—directly down the new path to and through the vegetable garden where the design ended. The view from the hill was the new grand finale of the garden.

Another year passed. Once again, I was back in the garden I knew well. Once again, Charles and I passed through the rooms and ascended the blooming hillside to the column. From there I gazed downward. Something caught my eye. There in the far distance was a prominent stone obelisk. Now a continuous visual line connected the hillside we were on to the axis of the vegetable garden and out across a gently rising pasture to the stately obelisk. A new far point had been created. The effect—of a landscape unified by design—was ennobling. It was exactly what a garden should be and what Charles Stick's gardens are.

May there be more of them in this world.

—Sam Abell

BEGINNINGS

THERE are few better examples of the imposition of order on a landscape than an aerial view of Chicago at night. From Lake Michigan west across the plains and prairie of the Midwest, a vast network of lights articulates the grid of structured, linear design interconnecting in every direction. Far below, within one of those tiny twinkling squares in a sprawling array of neighborhoods, is the childhood home of Charles J. Stick. Long before he understood how to structure the landscape, Stick experienced the innate value of order within nature.

"There was something magical about the street where I grew up. All the streets in Western Springs were lined by American elm trees. The space of those streets was defined by a cathedral ceiling of trees and that made a big impression on me. The visual geometry of the trees, the aesthetic tendency, the rhythm, and the space they created is still very deep-seated in my psyche."

The youngest child of Rex and Cecil Stick, Charles grew up in the suburbs of Chicago, where his first brush with managing a landscape was responsibility for the family lawn. Other forays into his future vocation were rooted in summer work at a nearby golf course and the vegetables and flowers he planted in the garden behind his parents' 1920s, Spanish colonial house, staking tomato plants with broken hockey sticks.

As a child, Stick was a budding hockey star. The son of a sporting goods representative, he spent much of his youth on ice skates, and he was invited to play for the Chicago Minor Hawks, a traveling all-star team of the most talented teens in the region. Stick's grades were "good enough," but his hockey skills garnered collegiate notice, and he set his sights on an Ivy League education. He applied to Harvard, Yale, and Princeton, but was not accepted. He also had considered smaller schools in the Northeast, including Middlebury, Colby, and Bowdoin, and during one of his hockey team's East Coast swings, Stick slipped up to visit Middlebury College, a small, private school in the Champlain Valley of western Vermont.

"As soon as I stepped on campus, I thought it was absolutely beautiful—set in an idyllic New England village set between the Adirondacks and the Green Mountains. When I came home, I told my parents I wanted to go to Middlebury. As Midwesterners, they just said, 'Well, where is that?'"

Middlebury did not offer athletic scholarships, and it was the second-most expensive school in the nation. Despite his parents and a skeptical high school guidance counselor suggesting other financially advantageous or proximate options, Stick remained determined to attend Middlebury and play hockey.

Hockey may have gotten Stick to college, but geology kept him there. The introductory geology course at Middlebury changed his mindset, his focus, and, consequently, his future.

Above: An aerial view of Chicago at night illustrates the geometric organization of the city and its suburbs.

Opposite: Elm trees arch over a street in Western Springs.

From left: Stick on the ice at Middlebury College; Brewster Baldwin, professor emeritus of geology; Robert Churchill, professor of geography.

"I absolutely fell in love with geology. And not only did I fall in love with pursuing it as my course of study, I began to admire the men who were teaching me; they became my mentors."

Professors Brewster Baldwin and Robert Churchill unlocked Stick's innate but untapped interest in the land, its formation, and, eventually, its storytelling potential. They simultaneously stimulated his love of learning through their enthusiastic and engaging classrooms and instructive field work. Stick immersed himself in his work—particularly Churchill's course in geomorphology—the basic study of land forms. "Everything on the land, the shape of the land has a story and all those things are connected," he notes. "For me, that started the notion of how the land tells stories."

Outside the classroom, Stick relished fieldwork with Baldwin, studying the bedding and sedimentation of the Champlain Valley, understanding the glaciers' movement between the Adirondack and Green mountain ranges and the reaction of the elements in response.

"As I'd drive through the valley, I'd wonder why the land was shaped like this. What made the rock do this? How does that stream move through that landscape? Now, I can't look at the landscape and not think about the underlying story. For me, geology emerged as a visual, almost intuitive pursuit of landform study and the basis of a scientific look at the natural world."

The Middlebury College campus is in the Champlain Valley, an hour south of Burlington between the Adirondacks and the Green Mountains.

To illustrate, Stick holds up a small stone, turning it over in his hand, its mineral components mysterious and yet definable. The rock has a story. Where did it come from? How did it get there? What happened to it along the way? Who were its neighbors?

"Gardens can tell stories just like this rock tells a story. My role as a landscape architect has always been as an interpreter. I'm trying to figure out what this rock and the land are telling me. With the people I'm working for, I'm trying to understand their story as well."

As Stick's undergraduate work continued, Baldwin and Churchill helped him develop an understanding and appreciation for what was in the ground and discover an awareness of the order above it.

"I went to Middlebury because it was so overwhelmingly beautiful, but I couldn't articulate what was beautiful about it at the beginning. There were some very handsome buildings on campus, designed in a classical vocabulary and constructed in 1801 in limestone quarried nearby. These buildings were growing out of the new nation and out of the landscape. I have a residual awareness of that."

Top: Middlebury College was established in 1800 in what was then a small town on the nation's frontier.

Above: Dan Kiley and Charles Stick at the University of Virginia after Kiley received the Thomas Jefferson Foundation Medal in Architecture in 1987.

The Middlebury experience resonated in an enduring way, connecting the structured order of his past and present, while hinting toward the future. That future definitively came into focus during his senior year when Stick met preeminent American landscape architect Dan Kiley.

"I was smitten by a young woman from Montana, and she asked if I wanted to go to a lecture in the dining hall by a landscape architect who was going to talk about redesigning the Middlebury campus."

Stick remembers Kiley as "an elfin character with his silver hair smashed up under his round glasses," suggesting some fairly radical design ideas that elicited wide-eyed stares from the students in attendance. Then he started talking about how to structure a landscape, about movement through the landscape. Kiley likened it to navigating down a ski slope. How does your body adjust to and react to the surroundings? What are the cues that move you through the landscape? What ties the space in between together? As an athlete, Stick understood.

"Even after I stopped playing hockey, I wanted to go to Holland and skate through the landscape on those frozen canals that just go on forever. How does one purposefully structure the landscape to make you want to move through it? Later, it became clear that it wasn't necessarily the physical body. It could be as simple as looking out a window and wondering how the landscape is structured. How does the eye move through the landscape? That was the seminal moment for me—when I met this

Above: The Lawn at the University of Virginia.

character, and the light went on. I decided I wanted to be a landscape architect."

Encouraged by Kiley and Professor Reuben Rainey, he enrolled in the University of Virginia School of Architecture. He's been in Charlottesville ever since.

GRADUATE SCHOOL

Thomas Jefferson's "academical village" is an enduring example of the architecture of a new nation, still fully visible with the iconic Rotunda as its architectural and academic anchor. Jefferson defined the American architectural vocabulary in the neoclassical tradition, imbuing it with the dignity and style of ancient civilizations in an effort to glorify its new place in the world. Stick recognized that the language of architecture had to be learned. "Looking around the University of Virginia, I knew I could spend the next three years learning this language."

At UVA, the graduate program in landscape architecture is part of the school of architecture, and students are required to have a foundational understanding of architectural principles as part of their graduate studies. "For me, it was a great advantage that architecture played a major role in my master's education. It formed the backbone of how I ended up looking at the world."

Without an undergraduate degree in architecture or urban planning, Stick enrolled in a three-month, summer "crash course" in drawing, composition, and design before his graduate studies formally began.

From the top: University of Virginia landscape architecture faculty: Ruben Rainey, Warren Byrd, and landscape architect Meade Palmer.

"It was an ambitious program to get people up to speed—many of whom had never expressed anything graphically in their entire lives," Stick remembers. "It was a little daunting, particularly learning to draw. I am fortunate to have been educated at a time when drawing—with a pencil in your hand—was still important. Those three months were a real awakening."

Stick credits Professor Warren Byrd for conveying both the importance of drawing and its capacity to unlock discovery. "He drew beautifully," Stick recalls. "He helped me appreciate the importance of drawing and how it influenced good design."

Immersed in the Beaux-Arts education at UVA, Stick studied the arts and architecture as the basis for the language of design.

To illustrate, Stick references Leonardo da Vinci's universally recognized Vitruvian Man, with its balanced proportions and its geometrical and philosophical correlation with the laws of nature.

"Our bodies are all part of that aesthetic development—the proportions relate to everything that man generates. It's the basis for the classical tradition—the original standard of design. Nature is the inspiration of all and 'man is nature.' Ultimately, the classical education is in search of those things. It supports a manifestation of those principles.

"The whole purpose of this exercise in garden making is to bring us closer to the natural world. To understand our place within it. Gardens are a manifestation of our desire to be a part of that."

Stick also was acquiring the practical tools to transform and amplify the landscape: plant identification; road design; surveying; site engineering; figure-ground design.

His undergraduate background in geology helped him appreciate the power of nature, while his graduate school studies offered a way to see how the built and natural environments could work together.

"One of the most interesting aspects of landscape architecture is the continuing dialogue between nature and art, and nature and architecture, and how they relate to one another—not just in terms of shade or shadow, but structures. The art of design is always about how those things relate to one another. How do you combine figure and ground to stimulate the eye in a positive way? Those are the sorts of things that lead you to draw a geometric figure on the board and think about the fact it has an energy that surrounds it and a relationship to whatever is juxtaposed against it. These are the building blocks of design—taking that two-dimensional idea and bringing it into three dimensions."

Layering those principles atop an understanding of the land provides the key to unlocking a site's storytelling potential.

Stick fully transitioned from outdoor undergraduate field work in the rugged mountains and arboreal forests of New England to the structured landscapes of the Old Dominion. While the language of design may be universal, it still requires skillful translation for practical application. Professors Rainey, Warren Byrd, Harry Porter, and Meade Palmer each imparted factual, functional, philosophical, and practicable lessons that shaped Stick personally and professionally.

In his history of landscape architecture course, Rainey emphasized the critical role of historic precedent in design and the collective integration of building and

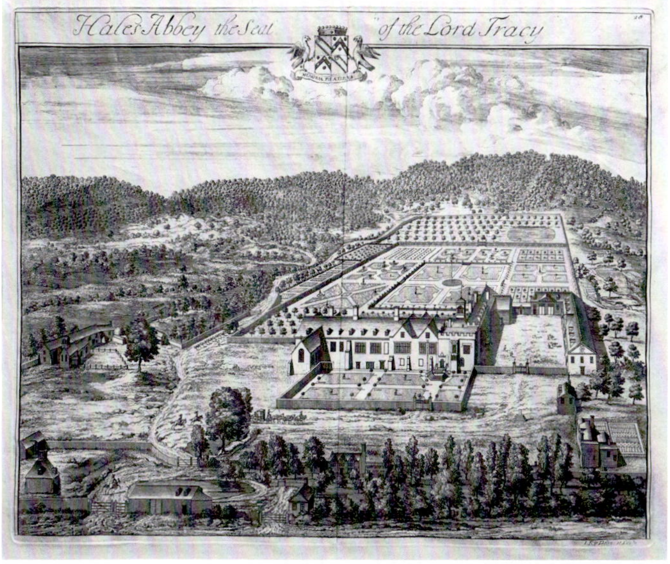

Left: Charles A. Platt, *Dutch Harbor*, 1883, from Stick's collection.

Above: Johannes Kip, *Hale's Abbey*, 1707, from Stick's collection.

landscape that ultimately would become the basis of Stick's practice. He also introduced students to the work of artist and architect Charles Platt, who enjoyed a successful career as an etcher and painter before achieving even greater acclaim as a designer. In 1894 a 33-year-old Platt published *Italian Gardens*, recording and interpreting Renaissance villas through his paintings, drawings, and photographs. After illustrating the fundamental integration of a house and its surrounding landscape—particularly the symmetry and ordered characteristics of his Italian subjects—Platt soon was commissioned to adapt those classical design principles in America.

"We were learning about organizational principles, composition, and horticulture. For me, Platt was someone who brought all those elements together. He was an artist, then an architect—almost by default—but entirely consumed with making sure that gardens were unified with the buildings he designed. Platt would design a house and work his way out into the garden with the Italian tradition of looking at the world and structuring the landscape with terraces and pathways. This notion of the integration of building and landscape were primary messages of my graduate education and would become the basis for my entire practice. It started with roots in the Villa Gamberaia and how that influenced Platt in his own work and then how his work influenced mine."

Before he could visit Platt's work or its Italian precedents, Stick took Rainey's advice and found examples closer to home in the great gardens of Virginia. Those site visits allowed him to experience classic gardens and the fundamental connection between house, garden, and the surrounding landscape in a tangible, active environment.

"Looking at gardens from the seventeenth and eighteenth centuries, it was clear that integration of house and garden are essential—from Platt back to Palladio and through history."

The massing of buildings and their spatial relationships once again resonated innately, while echoing similar principles gleaned from summers working in Kiley's practice: that human intervention upon nature is itself an act of nature. In *The Complete Works of America's Master Landscape Architect*, Kiley writes that "the greatest

Water parterre at Villa Gamberaia, an early twentieth-century addition to the original Renaissance garden.

contribution a designer can make is to link the human to the natural in such a way as to recall our fundamental place in the scheme of things."

With a growing artistic fluency in the language of design, Stick began to envision what kind of landscape architect he could be. While many landscape architects of the era pursued large-scale site development or urban planning, Stick chose a different path—one rooted in the examples from his studies of Platt and summers in Kiley's office. He was drawn to landscapes on a personal scale and fascinated by the ways houses and gardens relate to one another and the stories they could tell as a reflection of those who build and own them.

Ultimately, the model for his practice was another one of his professors: Meade Palmer. "He was such an inspiration to me—a country gentleman, a Virginian through and through, a foxhunter, a sportsman. He had a practice that allowed him to be involved in everything that went through his office. He seemed to be doing interesting projects, and horticulture was an important part of his practice. He loved plants. Many landscape architects don't talk about plants. They go into urban planning, and trees are just circles on a drawing. That wasn't enough for me."

Palmer's practice in Warrenton, Virginia, embodied a viable career on a scale that appealed to Stick.

"He was making gardens for people and, at the end of the day, that's what I have done. His life as a landscape architect was very appealing to me. He was very hands-on with everything he did, and his practice was extremely compact. If someone hired Meade Palmer, Meade Palmer showed up. I always thought that was the way it should be, and that's the way it's been with me. He showed me the niche he filled within the field was a real possibility, that I could make a living doing it, and it could be joyful."

Before he could join a firm, much less establish his own practice, Stick had to present his master's thesis, a year-long design project illustrating his understanding of the practical principles of landscape architecture and his own emerging design ethos.

Stick proposed a park on a long, narrow tract between the Chicago Water Tower and the shore of Lake Michigan. Bounded by Pearson Avenue to the north and Chicago Avenue to the south, the design is a testament to the new language he had learned and the respective dialects of those who taught it.

The Governor's Palace gardens at Colonial Williamsburg are a regional example of the connection between houses, gardens, and the surrounding landscape.

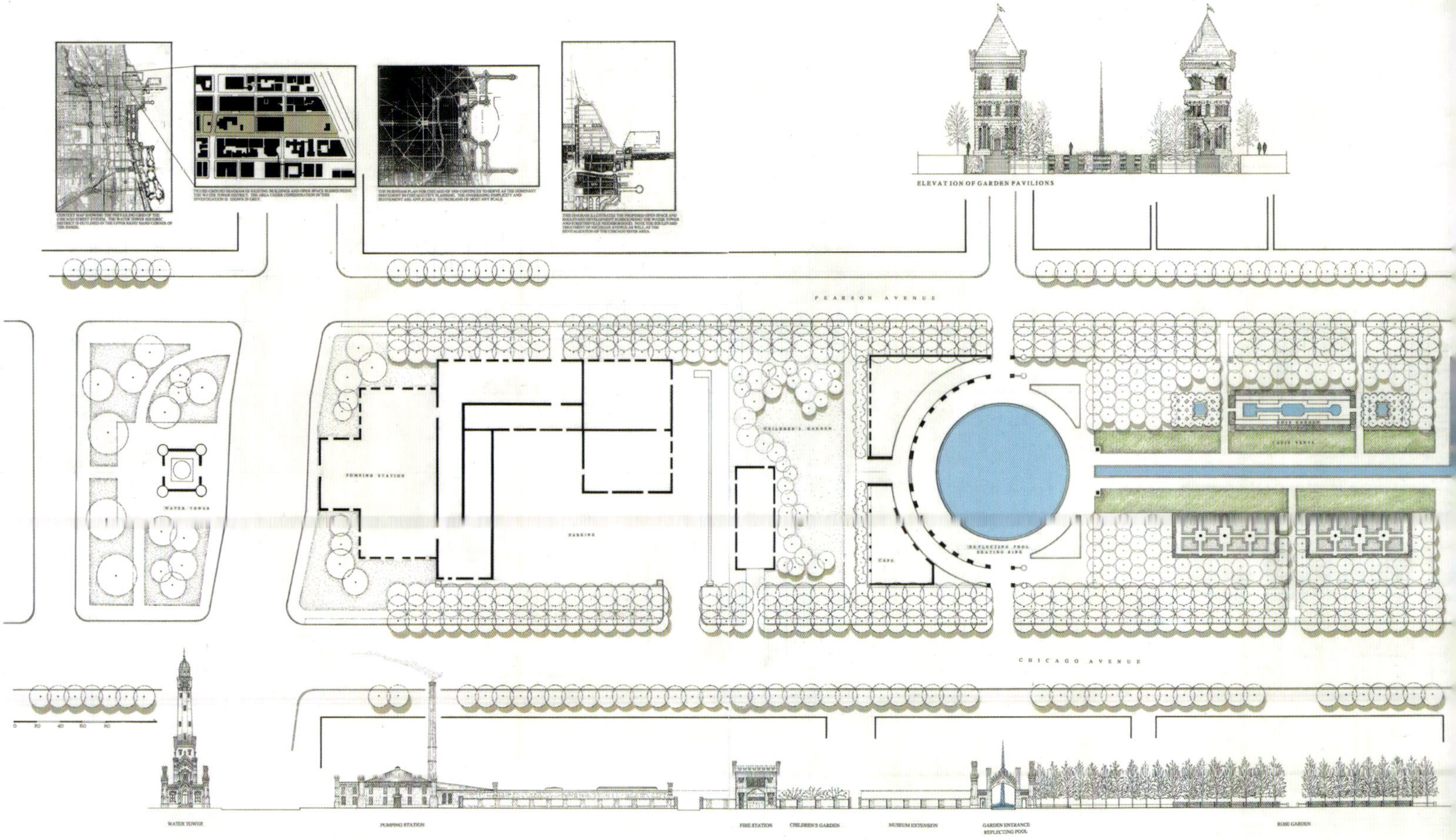

CHICAGO WATER

Stick's master's thesis project was a plan for the Chicago Water Tower Park that connected Lake Michigan and the city's water supply with the Chicago Water Tower.

"This is an experiment in my learning the language of design, much like writing a story in a foreign language," Stick says a little wistfully, flipping back through his old portfolio.

Anchored by the 1830s Gothic tower at the western end of the block, the design honors the city's historical water supply. From an intake at Lake Michigan, the water was pumped through an underground conduit between the two avenues to the five-story structure—one of very few to survive the Great Chicago Fire in 1871. The design celebrates water by linking the lake to the tower through a series of fountains, a water parterre, an ice maze, a cascade, and a reflecting pool.

"Dan Kiley's influence is clearly legible in that design," Stick explains. "The language that Dan spoke was a language I had learned well and felt comfortable within—a language of rigid control of plants and hardscape, of order."

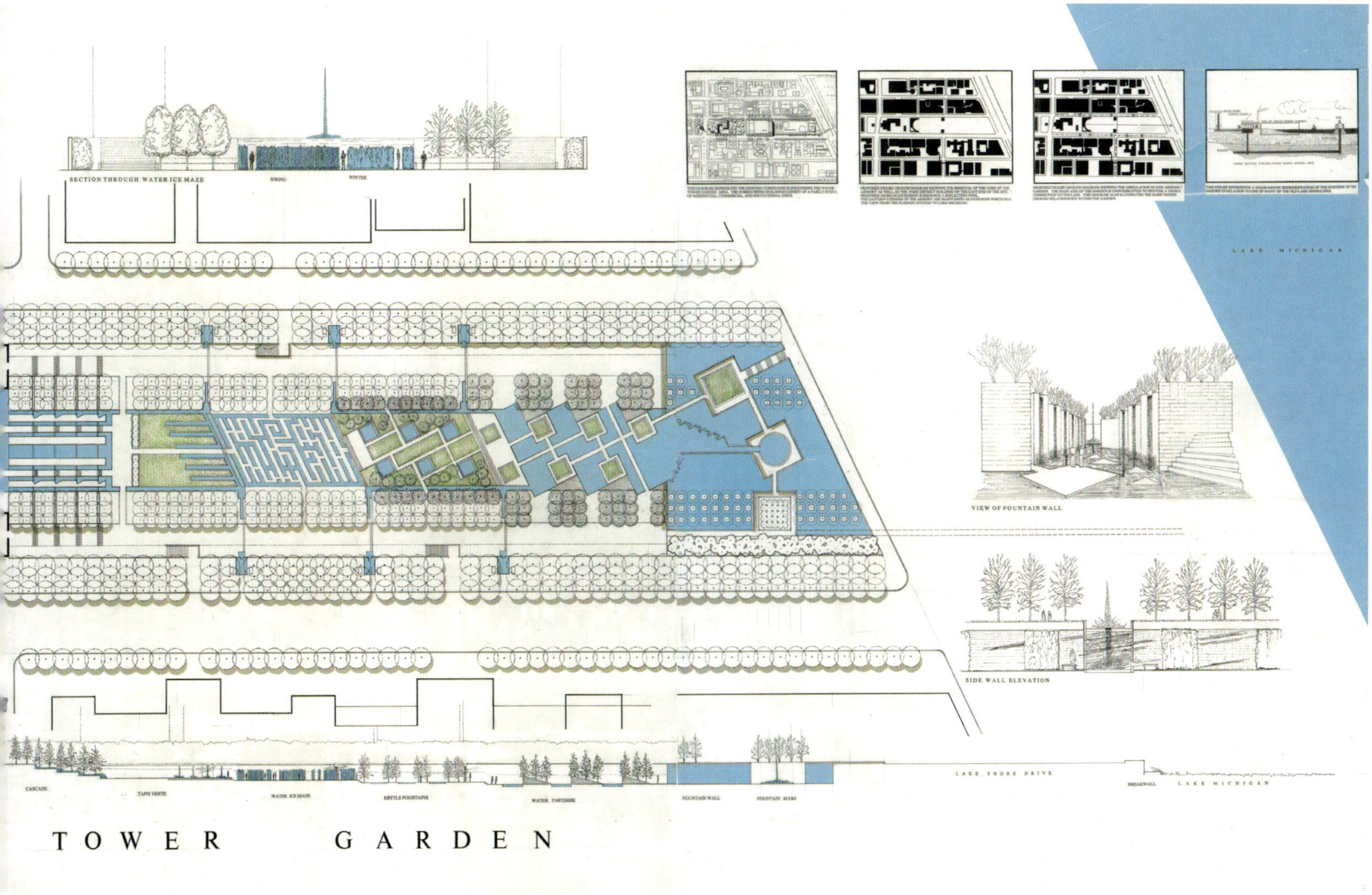

For Stick, the countless spoken and unspoken lessons in the language of design had crystallized into a coherent code. The elm-lined streets of his childhood marching him daily to school, the tenets of classical architecture at UVA, historical precedent, and the seamless integration of buildings and land were all merging together in the comprehensive and structured union of human administration on the landscape to unlock its storytelling capacity.

As Dan Kiley so often said, "Nature is the inspiration of all, and man is nature."

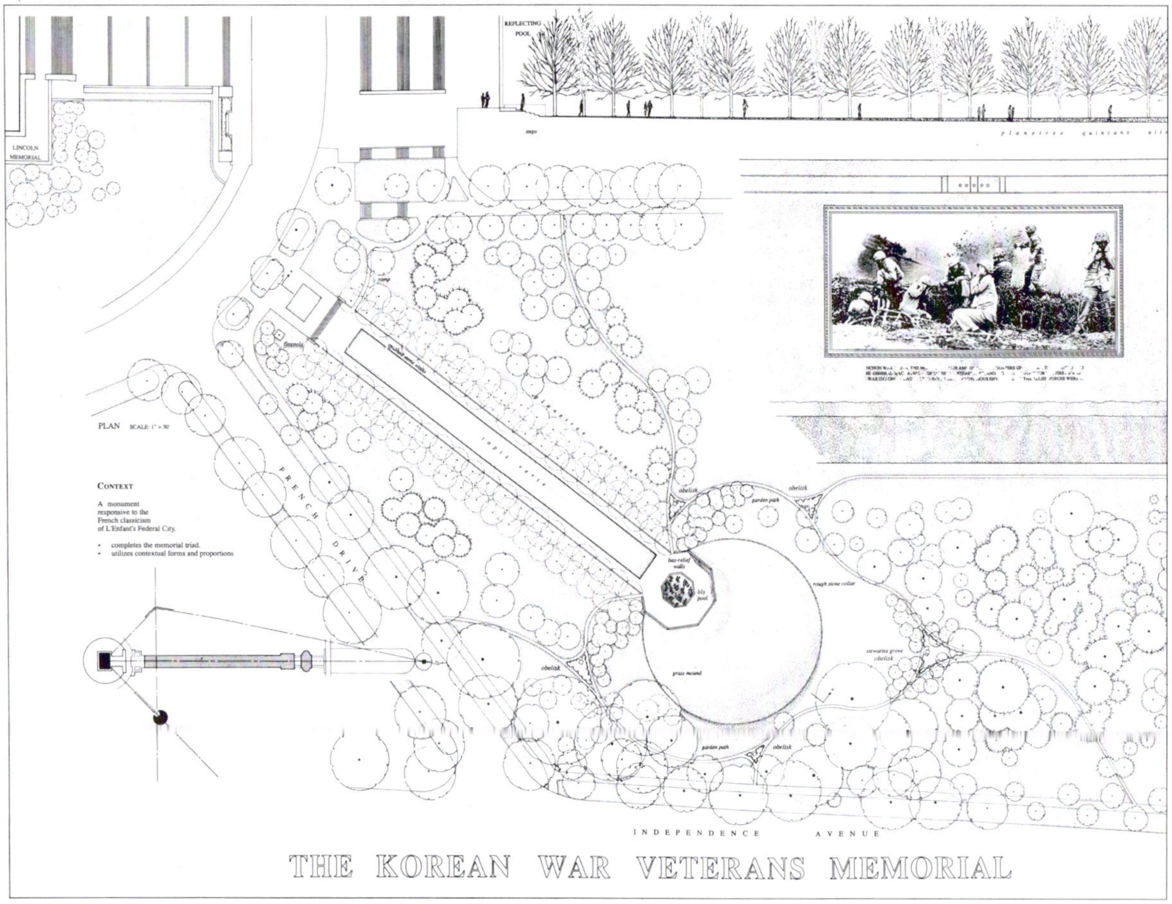
REFLECTING POOL
LINCOLN MEMORIAL
steps
PLAN SCALE: 1" = 30'
CONTEXT
A monument responsive to the French classicism of L'Enfant's Federal City.
• completes the memorial triad.
• utilizes contextual forms and proportions
FRENCH DRIVE
obelisk
garden path
bas-relief walls
lily pool
rough stone collar
stewartia grove obelisk
grass mound
INDEPENDENCE AVENUE
THE KOREAN WAR VETERANS MEMORIAL

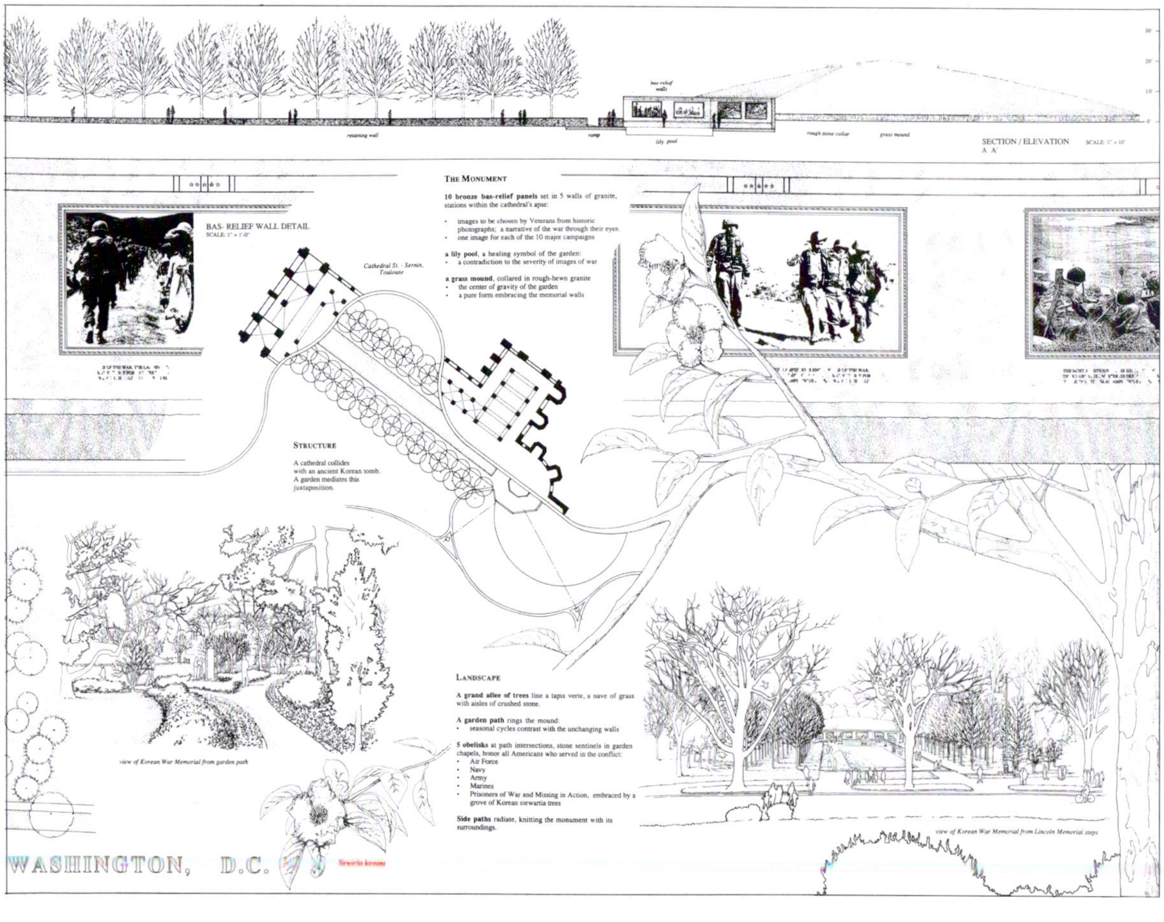
bas-relief walls
retaining wall
ramp
lily pool
rough stone collar
grass mound
SECTION / ELEVATION A A' SCALE: 1" = 10'
THE MONUMENT
10 bronze bas-relief panels set in 5 walls of granite, stations within the cathedral's apse:
• images to be chosen by Veterans from historic photographs; a narrative of the war through their eyes.
• one image for each of the 10 major campaigns
a lily pool, a healing symbol of the garden:
• a contradiction to the severity of images of war
a grass mound, collared in rough-hewn granite
• the center of gravity of the garden
• a pure form embracing the memorial walls
BAS- RELIEF WALL DETAIL SCALE: 1" = 1'-0"
Cathedral St. - Sernin, Toulouse
STRUCTURE
A cathedral collides with an ancient Korean tomb. A garden mediates this juxtaposition.
LANDSCAPE
A grand allee of trees line a tapis verte, a nave of grass with aisles of crushed stone.
A garden path rings the mound:
• seasonal cycles contrast with the unchanging walls
5 obelisks at path intersections, stone sentinels in garden chapels, honor all Americans who served in the conflict:
• Air Force
• Navy
• Army
• Marines
• Prisoners of War and Missing in Action, embraced by a grove of Korean stewartia trees
Side paths radiate, knitting the monument with its surroundings.
view of Korean War Memorial from garden path
view of Korean War Memorial from Lincoln Memorial steps
WASHINGTON, D.C.

PATHWAYS, PLACES, AND PEOPLE

"Good gardens are made of pathways and places."

AS much as Charles Stick intones that maxim, he could suggest that the principle has guided his life and career. Some paths are linear with clear objectives and outcomes. Others are more circuitous—their destinations anticipated but obscured. Some are entirely mysterious, yet ultimately rewarding. Like the paths in the gardens he creates, his personal paths have been practical and essential, connecting experiences and individuals with places and purpose. More often than not, the significance of a path becomes clear with perspective.

Those paths have brought him to gardens around the world and kept him comfortably close to his adopted Central Virginia home. They have enhanced personal and professional discovery and introduced profound and abiding friendships. Those paths have enriched Stick's life and enabled him to create something beautiful in the lives of others.

Stick's path began in the Midwest and would return there often, but by the time he finished his graduate studies, he was fully rooted in the classical architecture and gardens of Virginia. During his three years at UVA, he amassed a comprehensive blend of classroom instruction, field work, and entry-level experience, and he was eager to apply what he had learned. That opportunity emerged in 1988 in the national competition for the design of a Korean War memorial to be built on the National Mall in Washington, D.C. Stick and his colleague Peter Dunleavy collaborated on a design that earned an honorable mention out of more than five hundred submissions. Their entry featured a grass knoll, a reference to the traditional Korean burial mound. A tree-lined pathway led up and into the mound to an octagonal space featuring bas-relief sculptures representing the five major battles of the war. For young landscape architects looking to build their portfolios, the recognition was a valuable credential.

Above: Charles Stick in the garden house at Waverley.

Opposite: Stick and Peter Dunleavy collaborated on an entry for the Korean War Memorial design competition.

Subsequently, Stick joined the Belgian garden designer François Goffinet, who had opened an office in Charlottesville to design the grounds of Albemarle House, the six-hundred-acre estate of John and Patricia Kluge. Well-established in England, Goffinet positioned himself as a protégé of Russell Page. With Platt and Edwin Lutyens and Gertrude Jekyll, Russell Page embodied the principles of garden design to which Stick aspired. Looking back after more than three decades, Stick credits Goffinet for connecting him to many of the people and places that would shape his career.

Architect Russell Skinner first met Stick in Goffinet's office. For reasons neither he nor Stick can articulate, the self-proclaimed "odd couple" remain fierce friends to this day. "Charles, with his impeccable manners, good looks, and winning personality, was given a lot of responsibility early on, and François introduced him to clients," Skinner recalls. "Charles became friends with quite a few of them."

The Donald M. Kendall Sculpture Garden at the PepsiCo headquarters in Purchase, New York, was conceived by Russell Page.

As part of a stable of designers, Stick spent countless hours drawing the Kluge gardens and Carriage Museum, and he was eager to rise within the ranks of the firm. That opportunity appeared with the Donald M. Kendall Sculpture Gardens at the PepsiCo Global Headquarters in Purchase, New York, a project Goffinet took over after the death of the original designer, Russell Page.

In the late 1970s, Kendall, chair of PepsiCo, moved the corporate headquarters from Manhattan to Westchester County. In an effort to establish a modern corporate campus in a suburban community, Kendall proposed a public sculpture garden to complement the architecture by Edward Durrell Stone. The corporate offices are accommodated in seven square inverted ziggurats in the center of the property. The original landscape had been designed by the architect's son, E.D. Stone Jr., before Kendall invited Page to integrate dozens of sculptures into the carefully tended landscape.

"Russell Page developed a master plan for the overall layout and sited much of the sculpture Kendall had collected. It was a serious collection of modern, large-scale sculpture," Stick explains.

Page designed a connective "golden path" across the former polo field, linking lawns and specialty gardens, curving through woodlands of oak and birch and around ponds with large sculptures sited along the route.

The birch grove and Golden Path, designed by Russell Page, at the PepsiCo Global Headquarters.

"When Page died, the bones of the garden were laid out. Quite a bit of it had been implemented. For us, it was a matter of furnishing the various rooms and experiences that Page designed," Stick recalls.

By the time he died, Page had parlayed horticultural expertise and an artistic eye into a successful international career, becoming, in his own words, "the most famous garden designer you've never heard of."

"Page's style of garden making, and his extraordinary book *The Education of a Gardener,* continue to be tremendously influential," Stick says. "He was a consummate artist—a landscape designer with an artistic background. He also was an outstanding plantsman and influenced me to increase my horticultural knowledge."

As the Kluge project became more elaborate and consuming, Goffinet assigned Stick to manage the PepsiCo project, ultimately, if not inadvertently, positioning the young landscape architect on a path that soon would lead to his own practice. Meanwhile, accompanying Goffinet on a working visit to Leeds Castle in Kent, Stick had the opportunity to visit his first Page garden—and launch a lifetime of travel to the great gardens of the world.

"My very first morning in England, I woke up in the groom's cottage at Leeds Castle in the middle of a Russell Page garden. This was my first taste of England, and it was a wonder to me, an extraordinary experience. The notion of experiencing things firsthand cannot be overestimated."

"Waking up in that Page garden in England was one of those key moments," he continues. "It wasn't thought of as one of his great gardens, but it was all there—the structure and the horticultural richness within the planting beds. It never left me. That's where the love of travel—especially to the gardens of Europe—was fostered."

Above: The Culpeper Garden by Russell Page at Leeds Castle in Kent, England.

Opposite: Sketches from a trip to France in 1993.

As his work within the practice continued, Stick was starting to see a new and independent direction for his professional life. In addition to delegating the PepsiCo project in Westchester, Goffinet assigned him to help a new client, Nancy Gray, at her beach house on Jupiter Island, Florida.

"This was my first foray into working in a tropical climate. It was an eye-opening experience because of the environmental pressures on the horticultural aspect of the project." Stick would end up spending three months living on-site, guiding the project toward completion.

During the same period, John Bryan Jr., CEO of the Sara Lee Corporation, enlisted Goffinet to work at a David Adler–designed house on the North Shore of Chicago. Aware of his young associate's Midwestern roots, Goffinet invited Stick to join him on a site visit to the Bryans' property in Lake Bluff. Stick recalls Bryan—a corporate titan and philanthropist—articulating two specific tasks: to create a driveway to the house from Sheridan Road and to renovate the garden Ellen Biddle Shipman had designed at the front entrance to the main house.

"François had just gotten a big project for a movie producer in California, so after that meeting, he said to me, 'Charles, why don't you handle this project?'" Stick recalls. "I did and I never looked back. John Bryan ended up being a dear, dear friend and one of the great mentors of my life."

The various pathways of Stick's professional life were starting to intersect and expand.

LA PAGODE DE CHANTELOUPE ~ 25 MAY 1993
BASIN
PAGODE
BLOIS ~ THE CATHEDRAL
~ 24 MAY 1993.
25 MAY 1993
CHAMBORD

Above: John H. Bryan and Charles Stick in the garden at Waverley.

Opposite: A painting of Stick's office on historic Court Square in Charlottesville by Edward Thomas features "Squinty," the office cat, at the window. Inside, etchings by Charles A. Platt line the walls.

While Stick was implementing Goffinet's design at Nancy Gray's garden in Florida, she would introduce him to family and friends from Greenwich, Connecticut. During visits to the nearby PepsiCo site in Westchester, Stick often found himself invited to lunches or dinner parties within Gray's family and social circle, and some of those acquaintances began to ask him to consult on their houses and gardens. Those Greenwich connections included interior designer Isabelle Vanneck, who was working with retired satellite telecom executive Fred Landman at Sleepy Cat Farm, and Larry and Stephanie Flinn, whose garden at Cricket Hill would become Stick's first official independent project.

As his responsibilities and experience increased, Stick began to imagine what it might be like to direct his own practice. "I was managing the project at PepsiCo and flying down to Florida to work with Mrs. Gray. I was starting to have a presence in Greenwich, and I was going out to Chicago often and developing a deep friendship with John Bryan."

Before he could make the leap, he solicited Bryan's advice. "I flew out to Chicago and sat down with him on the lakeside porch of his house. We worked all day out in the garden and came back to the porch when we were done. I said, 'Mr. Bryan, I've decided to leave the office. Would you be supportive of that?' and he replied, 'Charles, you should have done it years ago.'"

Stick was determined to structure his nascent practice differently from the one he was leaving.

"There are hinge points in our lives when we perceive mistakes others are making. You decide you won't make those same mistakes. You will make other mistakes, but they won't be the same ones."

"François saw himself as the conductor of the people he hired to work for him. He brought in the horticulturist, the landscape architects, he brought in architects, and he formed the office."

Stick envisioned his practice differently.

"There is a phrase in the field: 'the art and craft of garden making.' I think it's a good phrase because the work is both an art and a craft. That's at the heart of what emerged in my view of the field."

Accordingly, Stick launched what began as—and has remained—a very small practice.

"The work of the last twenty years has been done with four associates," he marvels, citing architect Paul Gaertner, architect and renderer Peter Lorenzoni, architect Victor Colom, and personal assistant and office manager Pat Hotz. "It's never been more than that. I think it was a reaction to the hire-and-fire mentality across the field of architecture and a real testament to their devotion to the work we're doing."

With the support of Bryan and other clients, Stick and Meg King, an attorney working as an administrator in Goffinet's firm, formed a new business on Court Square in Charlottesville: Charles J. Stick, Landscape Architect. With King handling administrative matters, Stick focused on design and cultivating the business through his earlier projects.

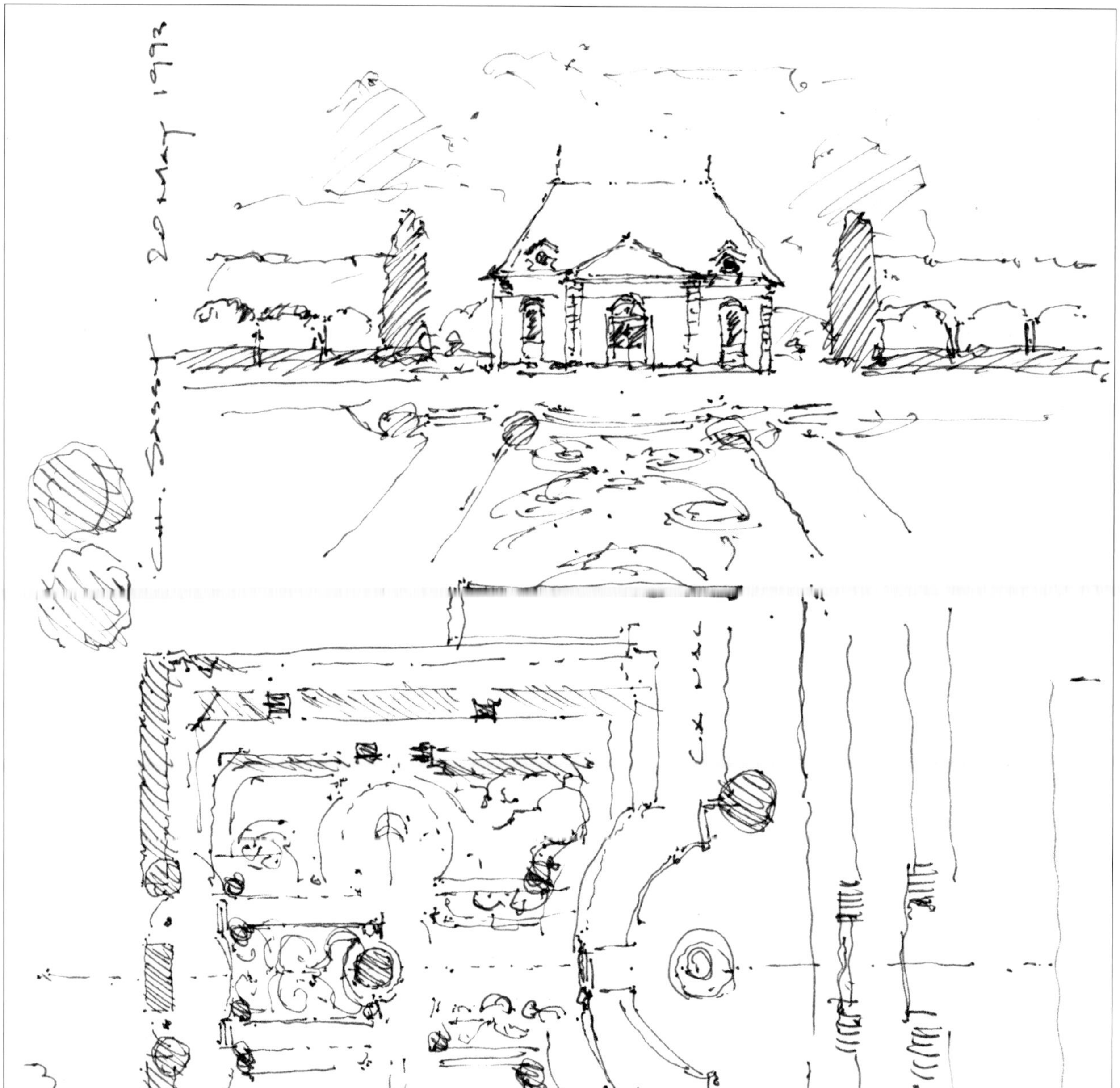
CH. SASSY · 20 MAY 1993
A GREAT JOY ON A RAINY AND COOL DAY IN NORMANDY

"My choice to keep the office small has defined how my practice unfolded. I always wanted to be the one doing the essential stuff—meeting with clients at their houses, working with them directly. I wanted to learn the stories of the people I worked for because those stories inform the decisions you make. Unless you know them, I don't think you can effectively make something good for them. In their finest and most interesting form, gardens can tell the story of the people who made them. That's evident in the gardens of Renaissance Italy where the symbolism and narrative are highly developed. Having a direct involvement with clients and their projects over an extended period of time is certainly the most rewarding aspect of it all."

Stick cites his time with Goffinet as a vital bridge between graduate school and his emergence in the broader field of landscape architecture. As he established himself in the profession, Stick forged enduring relationships with clients, contractors, and other professionals, some of whom he continues to work with today. In particular, he cites third-generation nurseryman Powers Taylor, who worked with Page on the PepsiCo project, and Long Island nurseryman and landscape contractor Mike Graham, whom he met working on a project in Southampton.

"I learned so much from them. They were nurserymen who genuinely knew plants."

He also credits Larry Labriola—who succeeded his father, Carmine, in the family landscape contracting business—as another influential figure, not only in the beginning of his professional development, but also in the years since.

"Larry worked on the PepsiCo project and became good friends with Russell Page—probably closer to Russell Page than anybody else in the U.S. Larry was there right from the beginning of my time at PepsiCo, and he's become a lifelong friend and colleague."

Above: Library in the garden house at Waverley.

Opposite: Sketches from a visit to Chateau Sassy in Normandy, 1993.

With the freedom to implement his own designs, Stick could blend his formal training with the influences of historical architects, designers, and horticulturalists whose work he admired. "The principles that define decisions in the world of architecture are very much the same as those we use to structure the landscape."

Opening an eighteenth-century edition of the Palladian canon, Stick says, "The world of classical architecture is within these covers. Much of the guesswork of architecture disappears if you have this as a guide. It's the touchstone. I want to use the same codified rules in structuring the landscape."

Stick had learned the organizational principles, composition, and horticulture of his craft and identified those architects and designers who would shape his art.

"Charles Platt was one of the people who brought all the elements together effectively. Lutyens was another who epitomized the integration of a house and garden, creating the envelope that allowed Jekyll to develop beautiful tapestries of plants within that framework. My career has been spent trying to achieve both."

In that pursuit, Stick prioritized continuing his education intellectually and experientially. Inspired by Professor Reuben Rainey's library, Stick has assembled a comprehensive collection of titles featuring the art and artists of landscape architecture. Meanwhile, he has researched, visited, toured, sketched, and photographed hundreds of gardens internationally, particularly across Europe.

From left: Hadrian's Villa, Chateau Brécy, the Capability Brown landscape at Chatsworth, and the Imperial Villa Shugakuin Rikyo, Kyoto, are among the gardens that have influenced Stick's designs.

"If you're going to invest yourself in this field, firsthand observation must be part of your education," Stick says frankly. "I've been fortunate that much of my life has included traveling around the world to see how other people solve problems. Is there something about the landscape that you're working on that allows you to call upon the traditions or principles of garden making to create a meaningful solution?"

For Stick, it goes back to the importance of curiosity: "Page's book has such a wonderful title: *The Education of a Gardener.* My travel and experience of these gardens has been the literal education of a gardener for me."

He also thinks back to viewing slides of the great gardens of the world in Rainey's class on the history of landscape architecture and the influence of Italian architecture professors Mario di Valmarana and Carlo Pelliccia, who brought their knowledge and experience from the Veneto to the Virginia Piedmont.

Rattling off an abbreviated history of garden making from the Italian Renaissance through the French baroque, and to eighteenth- and nineteenth-century English garden precedents, Stick cites the critical value of understanding and experiencing the history of his field. That knowledge is credibility currency with clients.

"I know I'm recommending the right solution for a property because it comes from building my own experience. I may not be the most knowledgeable person in the field or the most accomplished landscape architect, but I am invested in trying to do things based on historical precedent and informed by my experiences seeing how others resolve problems in the landscape."

He specifically cites Lutyens and his collaborations with Jekyll: "Their collaboration was a great union of architecture and horticulture that has had a lasting influence. Lutyens understood the seamless union of a house and a garden better than anyone else."

Referencing a seminal visit to The Deanery, Stick continues, "Looking at his plans, I could understand what he did to accomplish that. But in being there I experienced a richness, a joyfulness, and even a sense of humor in his work that you cannot feel unless you're there. He was willing to take chances, but at the same time, he was in total command of his craft."

Most people experience gardens as completed projects. The hardscape may structure the plot, while the plants tell the story. "One of the most difficult aspects in conceiving or maintaining a garden is teaching someone to see. There was no question that I learned to see through the people who have taught me and the people who inspired me, through their writings and their work. It's an essential piece of creating a garden and maintaining it at a high level."

After more than thirty years as a landscape architect, Stick has come to understand his focus and priorities:

"I've never claimed to be an engineer or social scientist. I'm not going to suggest a sophisticated anthropological site analysis. I'm knowledgeable about landforms and geomorphology and historical precedent. I wanted to design gardens for individuals. My goal has always been to work for people who were interested in creating gardens on a personal scale. I've always believed I was the fortunate one in that relationship between a client and a designer."

Skinner cites the special rapport his colleague and friend develops with his clients: "Charles wants things done absolutely right. His dedication to beauty—to creating something beautiful—is the most important thing, and he is absolutely committed to his work. For those who want that level of intimacy and connection between their gardens and their houses, Charles is their man. He attracts clients who are truly committed to their places because he cares so much and he's able to bring those clients along with him. And they just love him."

The gardens Stick makes are authentic: responsive to the land, reflective of the people making them, and respectful of the past. Stick invokes Vitruvius and his 2,000-year-old principles of good architecture: firmitas, utilitas, and venustas. "Firmness, commodity and delight or stability, utility and beauty," Stick translates. "That has always been a guiding light in my practice."

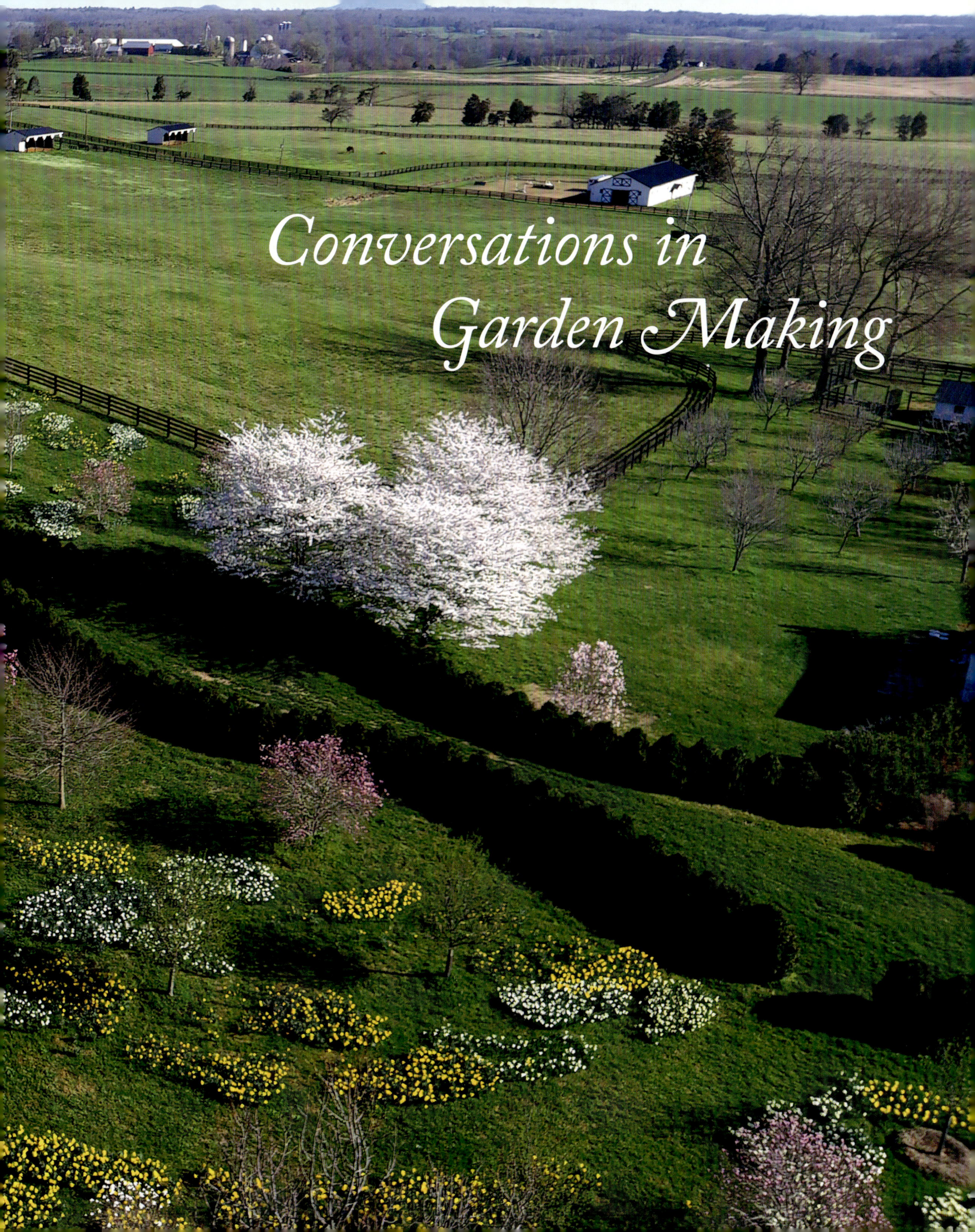
Conversations in
Garden Making

CRAB TREE FARM

LAKE BLUFF, ILLINOIS

LIFE may be defined by success and failure, but it is measured by transition. Age. Family. Career. Those periods of change from one condition into another are composed of foundational memories crystallized by opportunities, choices, experiences. More often than not, transitions are personal touchstones, with memories evoking powerful feelings of people, places, and purpose. Others may assess the outcomes of those key moments, but individually we each understand those critical junctures that shaped our decisions, ultimately defining who and what we are.

Charles Stick grew up just west of Chicago, but by the time he started his landscape architecture practice in 1994, Crab Tree Farm in Lake Bluff, north of the city on the shore of Lake Michigan, could have been his home. Over more than two decades, Stick concurrently cultivated a rich, woodland garden experience and a deep and enduring kinship with John Bryan and his family on their property.

Stick suggests Crab Tree Farm as a Georgian Arcadia, melding cultural and geographic eras to create a compound that encompasses the integration of English garden follies with the pastoral, peaceful beauty of the Peloponnesus in ancient Greece. "I think the spirit of Arcadia is transporting someone to another place in time, but also making a connection between the present and classical times," Stick says. "That's a major component of the notion of the Arcadian landscape, and I think that's what John and Neville Bryan and their family achieved at Crab Tree Farm."

In doing so, Stick suggests that Bryan simply wanted to create something beautiful and engaging. "The joy of experiencing the garden and the conversation that emerges from that experience is the genuine prize. That was his sense of inclusiveness and his desire—more striking in him than anybody else I've ever known—to expand other people's lives."

Above: A lead cello is a featured element within the walled garden.

Opposite: A path extends from the house through the Ellen Biddle Shipman garden to the Jefferson House, designed by David Adler in the 1920s.

CRAB TREE FARM HISTORY

Billed as the "only working farm on the shores of Lake Michigan," Crab Tree Farm traces its origins to a Judge Henry Blodgett, an abolitionist lawyer and lawmaker who relocated from New England. After his death in 1905, the property—then known as Ash Meadows Farm—was acquired by Scott Sloan Durand as a gift for his wife, Grace, a progressive dairy farmer, whose growing operation had become unwelcome within the residential Lake Forest community. Grace Durand transferred the name of her enterprise—Crab Tree Dairy—to the new farm.

By 1927, facing financial challenges and keen to develop the valuable lakefront property, the Durands sold a parcel to William McCormick Blair, whose great-grandfather Robert McCormick first tinkered with a mechanical reaper that was later patented by Blair's great-uncle, Cyrus McCormick. Prominent members of Chicago society, Blair and his wife, Helen, commissioned architect David Adler to design a summer home at Crab Tree Farm.

Adler was a prolific and respected architect in the Country Place Era before the Great Depression. An admirer and contemporary of Charles Platt, Adler deftly wove together architectural styles in a seamless way to create his own signature style. Adler exerted complete control over his projects—from custom millwork and details to reclaimed and historic interior design materials to integrated landscaping that complemented his houses.

At Crab Tree Farm, Adler constructed a shingled colonial revival house amid mature oak trees on a bluff overlooking Lake Michigan. With its moss-covered shingles, the gambrel-roofed house suggests it began life as a mid-Atlantic farmhouse before migrating to the Midwest. Normally a stickler for symmetry, Adler designed the Blairs' house with an offset H-plan suggesting that it had been added onto over time. The entrance is on the west facade where an enclosed garden by Ellen Biddle Shipman becomes a flower-filled foyer.

Opposite the garden, through a forested passage, Adler referenced the patriotism of the 1920s with the Jefferson House, a folly honoring the nation's third president and author of the Declaration of Independence. The small, classical structure is one of half a dozen on the Crab Tree campus west of the main house. These include a potting shed attached to a Lord and Burnham greenhouse, a series of caretakers' cottages, and a two-story glass-roofed tennis house, which gave the Blairs an escape from the city for invigorating indoor winter exercise.

The colonial revival house, designed by David Adler for William McCormick Blair, was completed in 1927.

A nineteenth-century limestone figure of Atlas stands on a pathway within the woodland garden that offers a distant view of Lake Michigan.

"Adler established the garden space in front of the house and put the axis through the woods to the Jefferson pavilion. He also was the one who arranged the location of the greenhouse and potting shed, and the caretakers' quarters," Stick explains, "His brilliance was in using the buildings to create garden spaces, not unlike the gardens in Williamsburg where the buildings also created garden spaces. The spaces in between can be utilitarian or pleasure grounds, and that's something I've always loved about Crab Tree Farm and Adler's genius."

Stick also admires Adler's skillful manipulation of the views. "You don't see the lake until you're at the house," he notes. "And that view is unimpeded." To the east, the distant lake horizon is unobscured by ornament or obstacle. To the south, the Chicago skyline is visible even on a hazy day. More subtly, the axis connecting the Jefferson pavilion to the Shipman garden bisects the main house, opening up a "secret" preview of the lakefront beyond. "I don't think this was an accident. Adler knew where his eye-level was, and he wanted to see the lake shimmering through the house from that point," Stick asserts. "Those are the sorts of things that Dan Kiley said may be serendipitous, but you have to take complete credit for creating such joyful experiences. Adler knew exactly what he was doing."

Working from Shipman's original plans, Stick renovated the garden at the base of the main house.

JOHN BRYAN JR.

Following William Blair's death in 1982, John Bryan Jr. and his wife, Neville, purchased Crab Tree Farm from his estate. John Bryan Jr. grew up in West Point, Mississippi, where one of the key local industries was the family-owned Bryan Foods, known as the "Flavor of the South." By the late 1960s, Nathan Cummings, CEO of Chicago-based Consolidated Foods, had added Bryan Foods to his portfolio and invited Bryan to join his administration. Focusing on diversification and decentralization, Bryan, as the new CEO, positioned the company for unprecedented growth by acquiring large companies with established brands while anticipating the potential of untapped international markets. In 1985 Consolidated Foods became Sara Lee Corporation, capitalizing on the company's high-profile, popular frozen-dessert brand.

While Bryan was transitioning Sara Lee into a prosperous international corporation, he also was preserving, renovating, and improving Crab Tree Farm. Unlike many estates of the 1920s, Crab Tree Farm has been preserved in its entirety. Initially, the Bryans employed Tennessee-based landscape architect Wiley T. Jones, who had worked for the family in Mississippi. Following Jones's death, a conversation between Bryan and then PepsiCo CEO Wayne Callaway led the Bryans to François Goffinet and ultimately the young, Chicago-area native within his practice.

"John Bryan was the greatest influence on my life," Stick says with sincere affection and admiration. "He was a mentor and, in a way, a father-figure to me. He represented how life can be lived graciously, sensitively, beautifully, fully, but also in service to others. It is obvious John would be remembered as the CEO of a large, successful corporation, but he's best remembered as a great, great philanthropist of Chicago."

Bryan raised hundreds of millions of dollars to support the creation of Chicago's Millennium Park, the addition of the Modern Wing of the Art Institute of Chicago, renovations to Chicago's Lyric Opera, and the preservation of the landmark Ludwig Mies van der Rohe Farnsworth House, among other Chicago-area projects.

At Crab Tree Farm, the Bryans restored the Solon Beman–designed barns that were the core of Durand's dairy operation, ultimately transforming them into galleries for world-renowned collections of English and American arts and crafts furniture and decorative arts.

"I have never known a more curious human being than John Bryan. I admired him so much because there was such a generosity of spirit, and my own curiosity was fostered just by being around him," Stick says. "I came to a point in my life where I realized that curiosity is one of the secrets of life. I remember him saying, 'It doesn't matter how you get there, it just matters how far you want to go.'"

John Bryan crossing the south pond.

Junipers and willows frame a barn on the working farm. Long-established honeysuckle and iris are planted along the banks of the north pond.

Bryan seemed willing to go as far as possible—amassing definitive, nearly legendary collections. He said he didn't have the resources to collect French Impressionist paintings, but he was able to collect in other fields. He assembled an extraordinary collection of Gustav Stickley furniture and Delft pottery and one of the most important collections of eighteenth-century English furniture and decorative arts in the country. For Bryan, it wasn't enough to have an object. He deeply understood it. He could account for its lineage, provenance, and purpose.

"The collection was integrated with the architecture of the Adler house; the decorative arts aspect of it was always layered to the fullest extent—the fabrics, the rugs, the brass," Stick notes. "It was all part of a comprehensive whole. The garden was a part of all of that."

THE GARDEN

Stick describes Crab Tree Farm as a woodland garden populated with follies and connected by a series of integrated pathways. The garden is an animate museum complementing the inanimate collection. Beneath a soaring canopy of century-old oak trees, an understory of maple leaf viburnum and gray stem dogwood creates a chest-high hedge along lush, grassy paths linking the garden follies and garden rooms in a considered series of vignettes.

"During the entire time I worked with John Bryan, I never lost the notion that everything was carefully considered—whether it was his collection of furniture or the decorative arts," Stick observes. "That devotion, that level of energy made its way out into the garden. It was all of one piece."

The garden features a hermit's cottage, an English pavilion, the Little House, a stone council ring, and a waterfall—all largely beneath the overarching woodland canopy.

"His knowledge and his curiosity were almost a backdrop for an understanding of the garden as defined by a series of buildings and engagements," Stick explains. "He had traveled and seen enough that he had a sense of how to create these epicenters within the structure of the garden. The scale of the garden allowed for that. It was a tremendous learning exercise for me—to be able to work on that scale with these garden structures and the hierarchy of how they all relate to one another. How do they reveal themselves in a meaningful and cohesive way?"

The Shipman garden adjacent to the main house.

Opposite: The hermit's cottage and the English pavilion are two of the follies featured in the garden. The hermit's cottage was built by hand and thatched with the help of Victor Chinnery; the eighteenth-century pavilion was acquired in Europe and relocated to Crab Tree Farm.

There was no master plan. The garden developed organically. It evolved specifically from Bryan's interests, his love of collecting, and a sense of creating delight. "As you make your way through the landscape, it's not only the joy of walking through this very beautiful woodland, it's also the building of engagement, of opening the door to one of the follies and entering another world," Stick explains. "That's what delight is about—transporting people to realms they have never experienced before. That's what traditional garden making is in the English landscaped garden—taking you to places you may not have expected or never knew."

The hermit's cottage, for instance, would have been expected in an eighteenth-century garden. It was a conceit. In the Georgian era, those who wanted something rustic or romantic created a grotto. Often, the hermit maintained the grotto and the hermit needed a place to live. At Crab Tree, the half-timbered, thatched hermit's cottage sits in a small clearing among soaring hardwood trees, surrounded by a rustic fence festooned with Virginia creeper and akebia. Inside, the fully furnished, one-room structure appears as if the occupant were out for the day caring for the grounds or perhaps trapping supper.

One of the first structures Bryan added was an English pavilion. He found the pavilion abroad and shipped it from Holland to Chicago. "The building has a definitive presence that required a relationship with one of the major buildings on the property," Stick explains. The obvious connection was with the tennis house and its strong classical vocabulary. The one-story, rusticated stucco pavilion anchors the south end of the property, flanked by a spring garden and a Chinese garden. Set against a backdrop of trees, the pavilion contains one of Bryan's collections of eighteenth-century English furniture.

A wide grass path extends to the north, connecting the pavilion to the tennis house and then continuing to join the original axis Adler created between the main house and the Jefferson House.

Blair had been a fan of Jefferson, but Bryan took the pedimented woodland folly to another level, installing specific Jeffersonian memorabilia. As a student at the University of Virginia business school, he certainly was familiar with Thomas Jefferson and Jeffersonian traditions. Between the Jefferson House and the main house, Bryan added the aptly named Little House. Constructed in 2000 of Wisconsin Lannon stone, the Little House contains a collection of three-quarter-scale eighteenth-century and Tudor-era furniture, complete with a monumental portrait of Henry VIII, surrounded by smaller likenesses of the monarch's wives.

"Once Bryan decided to create this building, we had to figure out how to site it," Stick explains. "The secret was to nestle it into the woodland without distracting from the main house or the tennis house. The purposeful scale of the building allowed it to settle gracefully into the landscape."

Additional follies, scattered across the land between Sheridan Road and the lakeshore, include a log cabin, a crow's nest, a tree house, and a trapper's cabin all connected in meaningful ways that do not distract from the main house and its radiating axis.

Stick's first task was the renovation and reconfiguration of the Shipman garden, a priority since it was so close to the house. The pair's next project was the walled garden, which was already underway when Stick became part of the Crab Tree Farm team. The Wisconsin limestone walls were in place, but no planting plan or interior structure had been developed.

Drawing inspiration from one of Bryan's Piet Mondrian paintings, Stick developed a geometrical planting pattern, creating an abstract thread to Bryan's arts and crafts collection at the Ellis House, which housed pieces of Harvey Ellis and Gustav Stickley furniture.

Together, Bryan and Stick viewed the walled garden as an outdoor gallery space. "It was a place for sculpture and plants to come together," Stick explains, and a tangible way to connect Bryan's collections directly with the garden.

Antique firebacks were integrated into the landscape design at the threshold of the rose garden and within the walled garden, initiating a practice of recasting antique utilitarian elements as exterior art. When Bryan began to acquire antique lead cisterns, and pieces of cisterns, Stick suggested installing them like paintings on the walls.

Situated between the rectangular vegetable and cutting garden and the circular spring garden, the walled garden is an oasis of lush grass, boxwood, and plants amid

Opposite, above: The tennis house was designed by James O'Connor, who often collaborated with David Adler.

Opposite: The Jefferson House in the oak woodland honors the nation's third president.

Above: The Little House contains a collection of eighteenth-century three-quarter-scale English oak furniture.

the ivy-covered limestone walls. Heavy doors enclose the sanctuary, preserving the peace within. The centerpiece of the garden is an eighteenth-century lead cistern on a three-step pedestal, filled with trays of sedum spilling over ornate carvings.

The nearby spring garden was the earliest use of the existing native plants on the property as the basis for the garden's architectural features.

"We agreed that we should encourage these native plants and use them to define the pathways of the garden," Stick explains. "Graystem dogwood and maple leaf viburnum are major native components of the understory. These masses are easily trimmed into hedges, and it helped reinforce connections between the buildings. The existing masses of mayapple are epic."

The Crab Tree Farm garden exists on multiple planes—the woodland canopy, the buildings, the understory—subsequently requiring connective elements to establish relevance, interest, and access.

The Adler design created the central connection linking the Jefferson pavilion through the main house to the lake beyond. From that pathway, the potting shed, greenhouse, and tennis house all are easily accessed. Paralleling that pedestrian path, the driveway skirts the northern property boundary, wrapping around the campus of caretaker cottages on its way to the barns and working farm portion of the property. Additional pathways through the garden connect the architectural features, establish a hierarchy within the garden structure, and introduce new vistas and garden experiences.

"Early on, John Bryan found a wonderful sculpture of Atlas and we needed to site him in a meaningful place," Stick notes. Using the tennis house as a backdrop, the two positioned the limestone statue facing east and carved a pathway through the understory, creating a forest-framed glimpse of the lake, granting the son of Titan a relaxing view amid his heavenly burdens.

From that path, a grassy swath unfolds toward the Little House like a large green welcome mat, acknowledging the building's presence in the landscape without trumpeting it. Farther east, the path joins another pathway that angles toward the main house in one direction and the treehouse in the other. Halfway between this nexus and the main house, Stick and Bryan designed and constructed a council ring, paying tribute to noted Midwest designer Jens Jensen with their interpretation of his trademark garden element.

Stick suggests that the council ring demonstrates Bryan's CEO spirit. Once Stick and Bryan decided to incorporate the stacked stone feature, translating ideas to action occurred quickly: "We had staked it out and measured it and gotten the feel for the bench height. Bryan told his mason to go ahead and build it, and the next time I came out, it was done."

"You're not going to find a raft of formal drawings for Crab Tree Farm in my archives," Stick continues. "There are some, but they're not as detailed as many others because a lot of it was dialogue, being on-site and that collaboration."

Other, more subtle pathways connect the main pathways with garden elements and sculptures, remote benches, and the beachfront below the bluff. Stick cites one path that he is fond of, which also illustrates the notion of connectivity, a key component of creating a woodland garden.

Opposite, above: A seventeenth-century lead cistern is a featured element within the walled garden. Framed by boxwood hedges, the cistern is planted with sedum; Boston ivy provides the backdrop on the Lannon stone wall.

Opposite: Designed in the style of Jens Jensen, the waterfall at the south pond is framed by masses of Sea Green juniper, weeping willows, and bald cypress.

A pathway cut through the west woodland frames the bell tower of the barns designed by Solon Beman.

"We were standing by the caretakers' cottages and looked southwest toward the farm. Visually, I thought it would be ideal to make a connection from the main house complex to the barns and the farm complex—particularly the bell tower," Stick explains. "John Bryan was very enthusiastic about the idea, and a month later, that pathway had been cut through the woodland yielding that vista. I think that was a particularly good stroke, and I learned clearly at Crab Tree Farm that so much of landscape design is about subtraction—particularly when you're working in a woodland garden."

Essential to the garden's construction, evolution, and maintenance was Bryan's active presence. "John Bryan was the eyes, the vision, the power of Crab Tree Farm. He powered it financially, aesthetically, and as a collector," Stick observes. "He was never a keen horticulturalist, but he had a sharp eye and was able to see the difference between the texture of this and that and understand from an artistic and aesthetic standpoint what belonged together and whether or not things had gotten out of bounds."

John Bryan, Neville Bryan, and John Bryan III.

For many years, Stick would fly to Chicago once a month and spend a weekend at Crab Tree Farm, working in the garden with Bryan and his staff, laying out pathways, siting a sculpture or building, planting thousands of bulbs and shrubs.

"There was always something going on—whether by the main house or out at the farm. I'd visit and the Little House was under construction, and we still needed to work on a pathway around the pond. There was a waterfall I had designed in the style of Jensen; it was all going on at the same time. He'd ask if we needed to let the stream come through or what we were going to plant along its edge, when were the daffodils going to emerge in the spring garden? So many projects were going on simultaneously."

During those regular visits, Stick came to appreciate the Bryans' teamwork in operating the farm and their individual contributions within the garden.

The farm's vegetable garden, flower garden, and greenhouse were all part of Neville Bryan's domain. Long before "farm-to-table" became a culinary catchphrase, Neville Bryan exemplified that philosophy.

"Neville has a deep gardening knowledge anchored in her southern roots and forged over a lifetime with her hands in the soil," Stick says. "Her gardening heritage was, and continues to be, an important part of her life and a genuine bond between us. I learned a tremendous amount from her from a horticultural standpoint. Her dedication to the spirit of this garden has been a great inspiration."

THE GARDEN IN TRANSITION

In 2023 Charles Stick returned to Crab Tree Farm to evaluate a garden in transition. By the time John Bryan died in 2018, the massive oak trees that created the protective leafy layer above the cultivated understory already had begun to decline. Historic rains relentlessly soak the impermeably dense, glacial clay soil, essentially drowning the trees where they stand. Five years ago, few others along the lakeshore seemed to notice the trees' plight. Now they all do.

Without that sunlight-buffering green veil, the shade-loving plants that comprised much of Stick's horticultural work in the garden are suffering. The more delicate ones wither and fry beneath an unfiltered sky. Others have exploded beyond control, crowding and obscuring the once-precise paths that connected the follies and features. Among the untamed, leggy wanderings of the native plants, enormous, ground-level stumps serve as horizontal tombstones marking where these protective giants once stood. Jagged, irregular openings to the sky above illustrate a garden and a property decidedly in transition. Yet, part of the garden still illustrates that which was: intensely lush and dark, dappled with just the right amount

Opposite, above: Atlas in winter.

Opposite: A water basin feeds the waterfall on the banks of the south pond.

of sunlight to nourish the understory, yet keep it in check. Here, the viburnum and dogwood coexist comfortably with the path—each seeming to understand its purpose, respectful of one another's boundaries. A birdhouse, just above the chest-high *Viburnum acerifolium*, appears as a wooden buoy floating upon a sea of green. The boughs of a dense forest arch above a grassy carpet and perfectly frame the bell tower on the old dairy barn off in the distance. A bench beckons down a lengthy, leafy path, suggesting mystery, contemplation, repose.

John Bryan III is committed to fulfilling his late father's charge: "Take care of Crab Tree Farm." He already has completed several book projects his father had undertaken before his death. But the trees and the garden present a more elusive challenge, and Stick has come "home" at his request to help.

Together, they walk the pathways, alternatingly offering observations and ideas, memories, and laments.

"Part of the garden had always been a little 'wild,'" Bryan acknowledges, "but I always felt like we had control of it. Now I feel like the plants have the upper hand."

He estimates the entire garden has lost approximately 40 percent of its woodland.

"At first my thoughts were, 'How do I stop it?' Now, I realize we can't."

What remains is to adapt.

Absent vigilance or presence, transitions occur that can be difficult to reconcile. People and places age and change, the forces of time drawing them further away from our idyllic, earlier memories.

"I almost feel like I'm in a place I don't even recognize," Stick admits. "It's shocking to see, but not daunting to tackle."

In addition to the mandate from his father, Bryan also must balance his mother's desire not to change things too drastically.

It would seem simple enough to convert the transitioning understory into a lush lawn similar to that which surrounds the house, but that does not suit the 88-year-old Neville Bryan: "I don't want it to look like a park!"

There is a consensus that the solution will focus on the paths and garden elements and attempting to restore the woodland. Stick and Bryan discuss amplifying the buildings and features within the landscape, elevating the garden's features while corralling the more aggressive elements amid the understory.

John Bryan III has any number of variables to consider in his stewardship of the property. Which trees will die and which ones will survive? For how long? How to protect the eroding lakefront and preserve the century-old buildings? Always and ever, it's a matter of maintenance.

"My father was more of an idea person—a builder. He was the visionary; I'm more of an operations person," Bryan explains, assessing the task before him. "Dad entrusted me because he knew I had the same enthusiasm he did, and I need to think about how the next generation can develop that level of enthusiasm—whether it's sustainability, low-maintenance, pollinator-friendly—whatever it is that interests them and gives them a reason to go out into the garden. I want them to want to come here."

Come they have. In the "summer house" the family built in 2011, all generations of Bryans bustle in and out in the waning days of summer. It's one of the few times

nieces, nephews, brothers, sisters, children great- and grand- all are able to assemble. Might there be an undiscovered gardener in the midst? Perhaps. Margaret French, the Bryans' daughter, is perpetuating her mother's horticultural heritage with her own efforts in the farm vegetable garden.

This multi-generational garden will continue to evolve, through addition, subtraction, natural and human intervention. It will attempt to balance duty with delight, re-creation with recreation. Bryan family members will find their own joy and delight, engaging the garden through their lens of beauty.

Above: John Bryan III and Stick consider the next phase of the garden.

Left: Pathways descend from the bluff to the shore of Lake Michigan amid existing understory of Staghorn Sumac and an overstory of lindens, poplars, and oaks, all pruned regularly to preserve the view.

For Stick, the delight is ever present, regardless of the challenge the garden may present: "My time at Crab Tree Farm was about one person who amplified my life. There I was, just going along, and all of a sudden, I meet this man and my life expands just like that—the learning, the appreciation I gained for everything just by being around him. On so many levels he was responsible for enriching my life by instilling a growing curiosity. Crab Tree Farm was and continues to be a great benchmark in my life, and now continues in collaboration with John Bryan III."

MOUNT SHARON

ORANGE COUNTY, VIRGINIA

FROM atop Mount Sharon—the second-highest point in Orange County—the Central Virginia landscape gently unfolds, seemingly unbounded in every direction. To the west, it is bordered only by the distant, inky smudge of the Blue Ridge Mountains and to the east, ultimately, only by the Atlantic Ocean. And to the southwest lies the inspiration, the connection, the origin and axis for this classic Piedmont property. Like many other elegant estates across the region, the house, the garden, the aura of Mount Sharon are inevitably (even if indirectly) connected to the region's most famous hilltop resident—Thomas Jefferson.

The Mount Sharon property dates back three centuries to a pair of eighteenth-century colonial land patents. Brothers John and Francis Taliaferro were awarded grants totaling more than 1,900 acres adjacent to nearby Bloomsbury, the prime, property of Colonel James Taylor II, one of the trailblazing Knights of the Golden Horseshoe and great-grandfather of Presidents James Madison and Zachary Taylor.

The first residence at Mount Sharon had been destroyed by the time Charles Champe Taliaferro inherited the land in 1876. In keeping with architectural fashion of the mid-1880s, Taliaferro constructed an ornate Second Empire–style house and enhanced the landscape, including adding the brick piers that still mark the entrance to the property. In 1935 the Taliaferro family sold the estate to Ellsworth H. Augustus, an industrialist and philanthropist from Cleveland, Ohio, and his wife, Elizabeth, whose sisters lived in Virginia. The couple commissioned a new house from New York architect L. Bancel La Farge, grandson of the painter and stained-glass artist John La Farge. The Georgian revival building features two gable-front wings flanking a two-story center block executed in Flemish bond with classic, dentiled gables and exterior molding.

Above: A pavilion modeled after an early twentieth-century garden building at Gunston Hall in Fredericksburg, Virginia.

Opposite: The arcaded rose arbor epitomizes the confluence of architecture and horticulture. *Rosa* 'New Dawn' and *Rosa* 'Zephrine Drouhin' climb the western red cedar structure while *Nepeta faassenii* 'Walker's Low' lines the bluestone walkway.

THE SEILHEIMERS AND STICK

Charles Seilheimer Jr., founder of Sotheby's International Realty, and his wife, Mary Lou, have lived at Mount Sharon for almost thirty years as stewards of the quintessential Georgian house that sits near the top of the hill amid centuries-old tulip poplars and American boxwood. Early on, the couple spent three summers touring the gardens of England, France, and Italy specifically studying the design concepts to enhance the grounds surrounding the house.

"We wanted a garden in keeping with the house—a union of the house and the garden," Charlie Seilheimer says. "I wanted a beautiful and significant garden," Mary Lou Seilheimer adds. "I was not interested in ornamental grasses or wild landscapes. I am not a plant collector. I'm not looking for exotics or rare plants. I'm looking for beauty."

Top: Mount Sharon was designed in the mid-1930s by New York architect Louis Bancel La Farge in the Georgian revival style.

Above: Mary Lou and Charlie Seilheimer.

An active, avid lifelong gardener, Mary Lou served on the Garden Club of Virginia committee that interviewed prospective candidates to succeed Rudy Favretti as its landscape architect position, one of whom was Charles Stick. He was not selected, but he was encouraged to interview for what ultimately became a collaborative design process at Mount Sharon. Stick was selected in large part on the recommendation of Favretti, who considered him a successor to the distinguished Richmond landscape architect Charles Gillette. As Seilheimer recalls, Favretti asserted, "There are others who over time will take his place, and one of them is somebody your wife already knows. His name is Charles Stick."

The Seilheimers selected Stick based not only on his portfolio and promise, but also on his personality. While Charlie Seilheimer brought an architectural and technical interest to the partnership, Mary Lou brought a lifetime of gardening experience. Stick brought the formal training and vision.

"My husband has some very definite ideas, and it can be off-putting to somebody who hears him," Mary Lou explains. "I needed to find somebody who could disagree with Charlie in a nice way. I could tell that Charles could do that. And we had a deal. If one of the three of us disagreed on something, we wouldn't do it."

Beyond the relationship dynamics, Stick needed to incorporate a diverse set of formal gardening requirements into the Mount Sharon design. Based on their European tours, they had compiled a list of key features: mixed perennial borders from England, massed color and formality from France, and architecturally structured garden rooms and fountains from Italy. They presented Stick with a list of mandatory and varied "ingredients" to include in a design that also would feature the historic boxwood.

"There's a phrase in the world of landscape architecture that developed in the eighteenth century in Europe and is attributed to Alexander Pope: 'The genius of the place,'" Stick notes, referencing the familiar phrase used to describe the universally big idea of a landscape. "For me, the genius of that place was the surrounding landscape."

"The house is beautifully sited by La Farge in that he resisted the temptation to build this house at the top of the hill," he continues. "The style of architecture of the house, the traditional layout of the rooms in relation to one another supports a connection between the house and the garden."

That harmony was still legible when Stick first arrived at Mount Sharon and identified the brick pathway making its way up the hill from the central axis of the house. Stick seized on that latent walkway as the bond connecting the house to the potential garden beyond.

Below: Mount Sharon is surrounded by 250-year-old tulip poplars (*Liriodendron tulipifera*). Brick gateposts mark an interior entry on the western approach.

Above: Design development drawing for the Garden of the Four Seasons.

Opposite: Master plan.

With the way Mount Sharon was sited originally—with each of its rear, first-story French doors opening to a level grass and brick terrace and its purposeful use of classical vocabulary—Stick envisioned seamlessly transitioning the interior structure of the house to its exterior surroundings.

"I wanted it to feel as though the spirit of the union of house and garden in a Palladian sense continued to ring true."

Stick maintains that the notion of good design is about a "big idea" and maintaining fidelity to the idea. At Mount Sharon, there was the house, a truncated walkway up the hill and, between surreptitious openings amid the burgeoning boxwood, the view of the Virginia countryside unfurling below.

Honoring all those elements, Stick also believes a good garden offers the opportunity for surprise and discovery.

"That was an important aspect of Mount Sharon's design for me," he explains. "I wanted it to be a series of discoveries, a series of focal points that moved visitors from one point to the next to discover something new. At the same time, it needed to all thread back to the house."

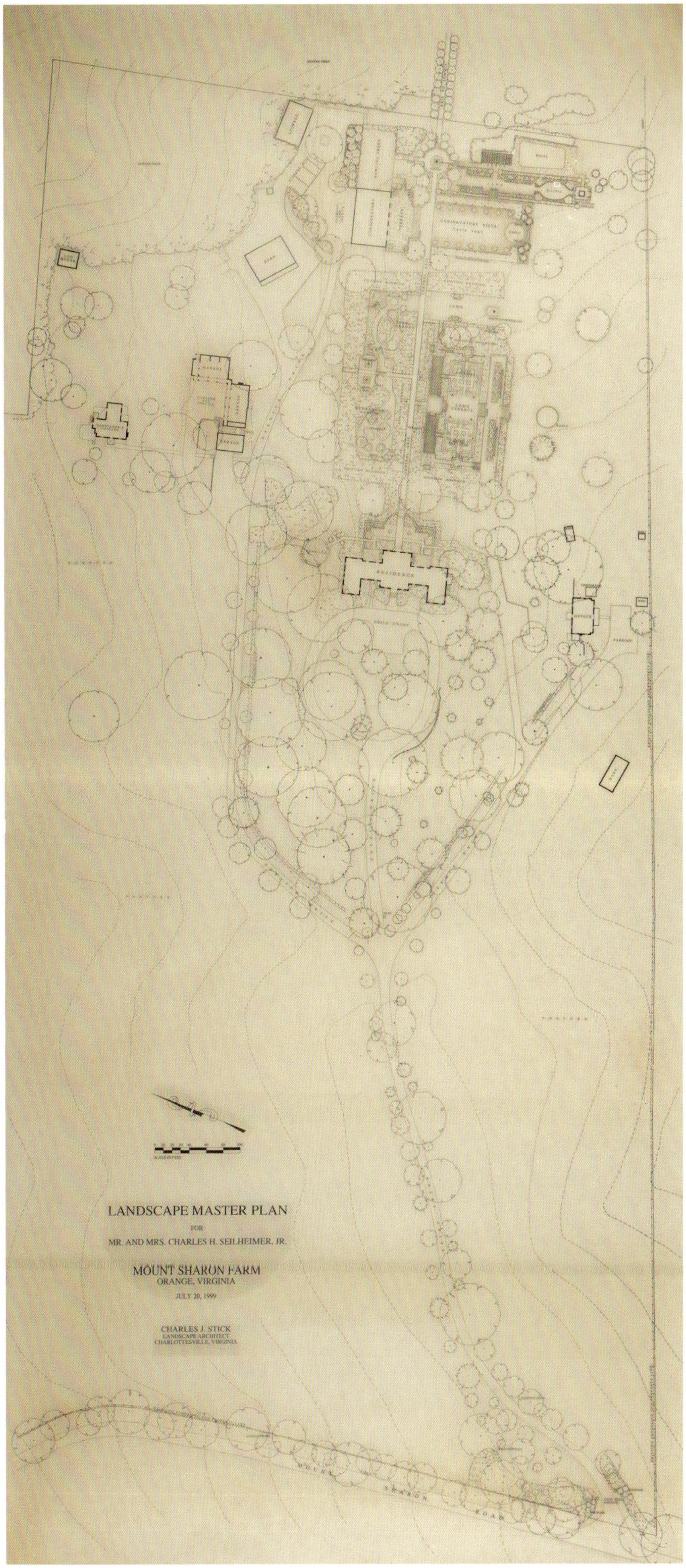
LANDSCAPE MASTER PLAN
FOR
MR. AND MRS. CHARLES H. SEILHEIMER, JR.
MOUNT SHARON FARM
ORANGE, VIRGINIA
JULY 20, 1999
CHARLES J. STICK
LANDSCAPE ARCHITECT
CHARLOTTESVILLE, VIRGINIA

Top: A fountain basin creates an intermediate focal point between the Garden of the Four Seasons and the surrounding pasturelands.

Above: A pre-construction photograph from the same vantage point. The garden required substantial excavation and infrastructure work by Duane Hawthorne and his team.

THE GARDEN

By the time visitors ascend the narrow, unlined rural road to Mount Sharon, they have already—if unknowingly—experienced the garden's greatest asset. It's only after they've entered the garden and looked upon the fields and furrows, plots and pastures, trees, and trundles that they can fully view and experience it completely. It may seem impossible to connect a limitless view to a finite footprint, but tulip poplars surrounding the house help visitors transition from the boundless to the bordered. Towering and impressive as the collective entrance to the garden may seem, it serves as prologue to the epic tale about to unfold.

Above, left: Under the direction of master mason Kendall Smith, crews install the bluestone walkway under the rose arbors above the Garden of the Four Seasons.

Above: The completed walkway beneath the rose arbor.

In the history of Mount Sharon, a nineteenth-century owner created terraced gardens long since lost to history, nature, and subsequent owners. When the Seilheimers purchased the property, only the boxwood suggested a planned formal garden design. Uninterested in a restoration or reclamation of any earlier gardens, the Seilheimers offered Stick an essentially blank canvas.

Across nearly ten acres, the Mount Sharon garden is composed of a series of distinct rooms opening off a central axis that begins at the rear of the house and extends up the hill to the excedra at the summit. Much like a house, what lies within each room—and the view from the windows—remains a mystery until the door is opened and a visitor steps inside. Certainly, part of the magic of the garden is that the panoramic view is both ever present and somehow exclusive.

"At every moment within this garden, I tried to open up the space to those distant views. No matter where you are in the garden, the main event reads clearly, and that is the view to the surrounding countryside," Stick says.

Stick's favorite part of the garden is the octagonal terrace, which is often overlooked in more formal summaries of the garden design. Just beyond the northwest corner of the house, a subtle pathway diverges from the main axis to open onto this terrace and a portrait of the Blue Ridge Mountains framed in tulip poplar. The subtle change in direction unlocks the majestic western view unavailable elsewhere in the garden and the octagonal shape reconciles multiple elements that converge in the space. "The way I look at the landscape is similar to the way traditional architects look at building; there are certain shapes that you're going to use to define those spaces," Stick explains. "Because of the angle at which you walk, the view, and relationship to the house, the octagon was the perfect shape to resolve all those things. When you step down into that terrace, it's all there in front of you—all of Virginia unfolds."

Standard floribunda and grandiflora roses fill parterres in the Garden of the Four Seasons. Low hedges of *Buxus* 'Green Gem' define the parterres, while Hicks yew (*Taxus x media* 'Hicksii') frame sections within the garden.

"One of the things that Charles believes, as I do as a classicist, is, that although you can have different adaptations, most of the forms have been tried and used before." Charlie says. "Most of the ideas that are worthwhile have been used, and you shouldn't be afraid to use them."

Here the original brick path became a passage extending from the central block of the house to create symmetry on both sides of the outdoor axis. From that fundamental classical element, each garden room exhibits its own purpose and atmosphere while maintaining a sense of continuity—not just to one another but also the house. That seemingly simple axis is the versatile element that connects the variety of garden styles and functions and bind those experiences together.

The garden of the four seasons is the most articulated, formal, and horticulturally rich experience within the Mount Sharon garden. In a series of formal, structured terraces, mirror-image rose gardens are separated by a green terrace with the figures of the four seasons presiding. Dozens of hybrid tea, grandiflora, and floribunda roses bloom amid boxwood parterres and pergolas. Subtle, centered fountains precede a large basin and jetting streams that direct the visitor's attention toward the boundless panoramic view.

Connecting all those elements, a soft-hued, two-part rose arbor spans an enchanted pathway paralleling the main axis and abutting the ever present boxwood. Stick deliberately divided the arbor to allow for a cross axis from the main path, leading down the terraces and out into the Virginia countryside.

On the opposite side of the path is the softly subtle hydrangea garden, complemented by the adjacent spring garden. Draped in the venerable boxwood, these rooms present a quiet sanctuary with lilacs, camellias, and other springtime staples selected and planted by Mary Lou Seilheimer. Curved lines throughout project a sense of comfort and welcome.

"I wanted the formality of the rose garden to contrast with a more relaxed arrangement of garden experiences, but I wanted them to be threaded together in a sensible way," Stick says. "Certainly the main path provides an essential continuity in itself, yet it also allows each room to have its own identity."

Massings of *Spiraea nipponica* 'Snowmound,' *Deutzia gracilis*, rhododendron, azalea, and camellia border the spring garden, offering an early burst of color amid the old boxwood hedges.

Emerging from the spring garden, and farther along the path, a large, level lawn extends deceptively into the infinite countryside below. A monumental urn provides a focal point and a measure of scale to the ethereal expanse beyond. Teased with that view, the visitor heeds the call of Eros at the far end of the axis and heads farther east along the path and up the hill to find a tapis vert, initially defined by pleached hornbeams, funneling the visitor's gaze toward a circular basin. A solitary jet of water rises from the basin, which is framed on either side by a second layer of arborvitae, layering the view both horizontally and vertically, but never obscuring the essential relationship to the Virginia countryside.

While the hydrangea and spring garden sight lines are defined by their gentle, understated curves, the orthogonal discipline of the tapis vert points directly toward the southeast horizon. Above the tapis vert, a mixed herbaceous border culminates in an elliptical garden room, featuring a white latticed pavilion originally built for the Seilheimers' property in Warrenton and moved to the site. The transparency of the structure creates a resting place to enjoy the view without disrupting it.

From here, at the garden's most distant moment, a cooling updraft rises from the fields below and allows the visitor to take in the entire landscape, with the main house

Pleached hornbeams (*Carpinus betulus*) line the tapis vert leading to a circular basin framed by an apse of Green Giant arborvitae. A solitary jet adds a vertical focal point to the garden room.

In late spring, the herbaceous border is alive with color as roses bloom on tuteurs amid foxglove, catmint, and lady's mantle. Wisteria covers a teak arbor beyond.

partially visible in the distance—an unexpected, yet tangible, anchoring discovery. Much as the distinctive dome of Monticello is intermittently visible throughout the surrounding gardens and grounds, Stick specifically sought opportunities at Mount Sharon to connect the garden experience back to the house.

"It was extremely important for me when we designed the garden, that there were places where that reference was evident," he explains. "What's interesting about the house is that you never view it all at once—even when you're driving up to it, because of the tulip poplar trees. That's a lesson in what the trees can do to give the house scale."

Building the Mount Sharon garden on the unstructured grounds surrounding the house required massive earth-moving efforts to level lawns and rooms, carve out terraces, and dig basins. Stick specifically cites the expertise of Kendall Smith and Duane Hawthorne. "Duane did all the grading and drainage work. If I could take him with me across the country to all my projects, I'd be much better off," Stick says. "And Kendall takes the long view in terms of construction. He and his team are not going to build anything fleeting."

The Mount Sharon garden experience is much like attending and enjoying a stage performance and—without being able to pinpoint what exactly it was that made it

Above: Green Gem boxwood hedges flank opposing fountain basins on either end of the rose arbor. The low hedges create the illusion that the water is floating on a boxwood plinth.

Opposite: Green Gem boxwood also lines parterres and stairways descending from the rose arbor through the Garden of the Four Seasons. The slopes are planted with a groundcover of *Liriope muscari* 'Big Blue.'

amazing—you simply know that it is. The writing. The music. The set. The acting. The direction. The details. Individually and collectively, the elements generate the sense of awe and create an indelible memory.

"Hopefully, when people visit Mount Sharon, they feel that this is a garden that is tied to a house, and those two things are tied to this landscape. That was the big idea. And I hope that big idea is legible, and I want them to have joy in what they're experiencing," Stick says. "What I've learned in visiting gardens across the world is that the art of garden making allows for an expression of personal stories. I try to give my clients a sincere and solid structure, a good framework, and often in the midst of that framework, their own pictures emerge to amplify anything I could have imagined. My wish is to help people make the gardens they want in their lives."

"I love being in here," confirms Mary Lou, who daily works in the garden, trimming, pruning, planting, making lists of tomorrow's tasks. "We spend a lot of every day being grateful and appreciative and just experiencing the joy and the beauty in this garden. It's absolutely a happy place, a very special place."

"It's been the joy of our life to have these gardens," Charlie Seilheimer concludes.

SLEEPY CAT FARM

GREENWICH, CONNECTICUT

TIME.

Even more than the soil, the cool Connecticut air, and the ever-arcing sun, time may be the most essential element of the garden at Sleepy Cat Farm.

Pipe clenched between his teeth, hands clasped behind the tails of his tweed jacket, Charles Stick walks through the garden he and Fred Landman began in 1997 and deferentially invokes the patron saint of mature plantings and seamless design. Time.

On the surface—or across the garden's thirteen acres—his reference is to plant growth and development, the organized achievement of the human influence on a disordered landscape. The fragrant viburnum, the redbuds, the wild geraniums blanketing Clapboard Ridge, hearty mountain laurel, the complementary understory of ferns and hostas coalesces to illustrate the collective benefit of time. But Stick's meaning is not exclusive to the overt horticultural prosperity. He is also talking about his gardening relationship, friendship, and trust with Landman, which runs as deeply as the owner's commitment to his plant-filled paradise.

It nourishes absent nutrients. It fortifies without food. It endures and evolves, ever adapting to its environment. Time.

It's been more than a month since Stick travelled from Central Virginia to the southern Connecticut to visit Landman and his wife, Seen Lippert, and the garden they have created. Since then, the cherry trees have transitioned from their debutante pink to a more subtle, everyday green. Tens of thousands of daffodil blossoms have bloomed and faded, and blindingly bright snowdrops and iridescent bluebells have emerged to take their place. Cold-frame lettuce has since been harvested for salads, and last month's vegetable seedlings are now sheltered by the dwarf boxwood hedges bordering the beds of the kitchen garden. The property glows with the unmistakable radiance of new growth in that color reserved exclusively for spring.

On a damp late April afternoon, Stick walks the garden from its lowest point back toward its farthest end in search of its owner—rumored first here, then there—an omnipresent apparition amid the plantings. Stick and Landman find one another, fittingly, near the Spirit Walk in the heart of the place and fall into that language time alone translates. Small talk is superfluous. The real issue is the garden and what needs to be done.

Landman is all about what needs to be done—one hand firmly gripped around a cluster of uprooted garlic mustard weed, its peppery aroma floating in the cool

Above: Azaleas, Virginia bluebells, and narcissus line the Golden Path.

Opposite: The Chinese Pavilion stands out amid a veil of *Magnolia* 'Dr. Merrill' and *Prunus x subhirtella* 'Pendula' in early spring.

Above: A moss-covered walkway planted with spring bulbs, narcissus and spring phlox connects the main lawn with the Golden Path.

air as he gestures hither and yon with garden problems and progress. Stick nods knowingly as Landman, ever the CEO, rattles off a torrent of tasks. As they traverse the garden, Landman and Stick observe, critique, marvel, and comprehend without further explanation. It seems incomprehensible, but it's almost as if they know each flower, tree, or shrub individually. It's quite possible they do. They created this, but nature and time made it what it is.

The Sleepy Cat Farm garden isn't magical—because magic is an illusion and the garden is very, very real. Instead, it could best be described as mystical and mysterious—each mossy stone path or straight-edged hedge offering opportunities for discovery and delight. On an already damp April morning, another spring shower adds an ethereal layer to the verdant richness.

Stick consistently contends good gardens are made of pathways and places, and Sleepy Cat Farm may be the best illustration of that maxim. There's the definitive Golden Path. A Chinese pavilion. A pebble terrace. A woodland garden. A berry wedge and fruit orchard. A vegetable garden and greenhouse. Prayer wheels. Grassy lawns. A perennial border. A meditation walk. A cascade. A grotto. A parterre. A cat maze. A wisteria arbor. So many distinct places, yet seamlessly linked by subtle, inviting pathways. It could be overwhelming, and it can be—from a sensory standpoint—but not physically or philosophically. The various elements of the garden, many of which are paeans to the great gardens of England, work in concert with one another. Like movements in music, each part can be appreciated for its distinct moment and for its critical role in creating the whole. For these gardening partners, Sleepy Cat Farm is an enduring, infinite garden experience.

Top: Fred Landman and Seen Lippert.

Above: Landman and Stick in the garden at Sleepy Cat Farm.

FRED LANDMAN

You could say Fred Landman has lived two lives. Perhaps three. However many it may be, they each are hallmarked by hard-earned success forged through collaboration and trust.

In his "first" life, Landman was an executive in the growing Spanish International Network (that would later become the global multimedia conglomerate Univision). In the mid-1980s, Landman was asked to direct a nascent private satellite company—PanAmSat—to broadcast television internationally, delivering emerging cable TV networks such as CNN and ESPN to untapped markets. By the mid-1990s, PanAmSat merged with Hughes Communications. Landman directed PanAmSat's global operations, sales, marketing, strategic development, and international affairs. He retired in 1999 at the age of 53.

The creation, development, maintenance, and future of Sleepy Cat Farm could be considered Landman's second career. He purchased the property and the substantial if dated Georgian-style house in 1994, five years before his retirement. Many of the characteristics that served him well in business have translated to his ongoing role as CEO of the Sleepy Cat Farm garden. He remains an outcomes-based executive, who trusts his team, prefers to listen rather than talk, and is personally invested in the challenge at hand. Naturally, a primary difference is his current board of directors—the farm's namesakes that arrived when he and Lippert married in 2001—are more interested in sun-dappled chair cushions, perpetual access behind closed doors, attention only on their terms, and, of course, food.

With a quick wit and a wry smile, Landman, like the cats patrolling the farm, seems to understand a secret that others do not. Maybe it's something he found in the garden or in one of his private meditations. Clearly, there's an inner fire and peace in the man that don't so much compete as complement.

He's quick to credit his team—citing Stick, architect Chuck Hilton, interior designer Isabelle Vanneck, horticulturalist Alan Gorkin, master mason Mauro Fidelo, and landscape contractor Larry Labriola.

"This is not my creation. These are the people who made this," he says. "I just happened to be the guy who said 'Yeah, let's do it.'"

"You learn a lot by listening to smart people," Landman continues. "What kills me is people who hire someone who is really good in their field and then they tell them what to do and how to do it."

You might say Landman's third life began during Stick's fiftieth birthday trip to Tuscany, when a bicycle ride through the country ended with Landman waking up a month later in an Italian hospital, weak, bewildered, and recovering from an emergency heart transplant.

"I thought, 'What's going on here? Where am I?'" he recalls. "I'm a very lucky guy. That was a strange alignment of stars that made that all happen."

Regardless, each experience has been rewarding in its own way. Asked today how he'd identify himself, Landman somewhat coyly responds, "I'm a gardener."

THE GARDEN

When Landman acquired the property in 1994, the house received most of the attention. Chuck Hilton, a young local architect, reimagined and renovated the interior and designed exterior additions to achieve the traditional symmetry associated with classical Georgian design. Landman, meanwhile, painted the first gardening strokes at the property with a small vegetable garden, flowers, and a neoclassical tool shed at the front of the property. But as Hilton moved on to reconfigure the uninspiring swimming pool and adjacent pavilion, Landman recognized he needed to expand his team.

Isabelle Vanneck, Landman's interior designer, recommended Stick, who had been working for friends of hers. Of Landman and their initial encounter Stick recalls, "We had a nice chat and walk around, and he asked how we could go forward. I told him he needed an overall plan for the property and within that he could decide what to focus on, and he said, 'Well, why not!'"

"I had the time. I had the funds. I was retired, and my work ethic is to do something, to make something, and that's what we did," Landman says. "I wanted to do something that I really loved. And I wanted to do it to the best of my ability. That's basically what I've been doing for the past twenty-six years."

As the house took shape, it not only needed to connect with the land, but also be received by it. Stick and Landman agree that landscaping often is deferred or simply dismissed because of higher-than-expected construction costs. Consequently, the house and the land are out of tune. Even before they began working and traveling together, Stick and Landman had visited great gardens of the world and understood—even if from different vantage points—how a house and garden work in concert with one another.

"One of the things that's special about our professional relationship is I've always looked at it in terms of Charles Platt's career. He would tell his clients, 'I'm building you an Italian Renaissance house or a house that has roots in that tradition. I think it's important for you to see what that means in its native environment, to see why it functions the way it does,'" Stick says. "His clients would go to Europe and spend time understanding those building traditions. If you had money and the time and were building that kind of house, it was important for you to understand what your architect was doing and why. Fred bought into that. He took me up many times on those suggestions, and eventually we began traveling to those places together."

Landman notes, "We didn't try to make a French garden or an English garden or an Asian garden. We tried to make a garden experience—a place where people go and find pathways that will lead them wherever they want to go."

Those pathways and places should lead to discovery, or, as Landman suggests, "surprise."

"You come around a corner and it's just a sense of wonder when you walk through this garden. You just don't know what will come next."

But before they could create those discoveries, they had some more tangible tasks to tackle—most notably the area immediately surrounding the house. They first reconciled

Opposite, top and right: The entry court circle is defined by a closely cropped *Taxus media* 'Densiformis' hedge and globes of American boxwood sent north from Virginia. "F-hole" borders of boxwood enhance the lawn.

Opposite, left: The bluestone pool terrace is planted with Tardiva hydrangeas and pots of Endless Summer Blue hydrangeas.

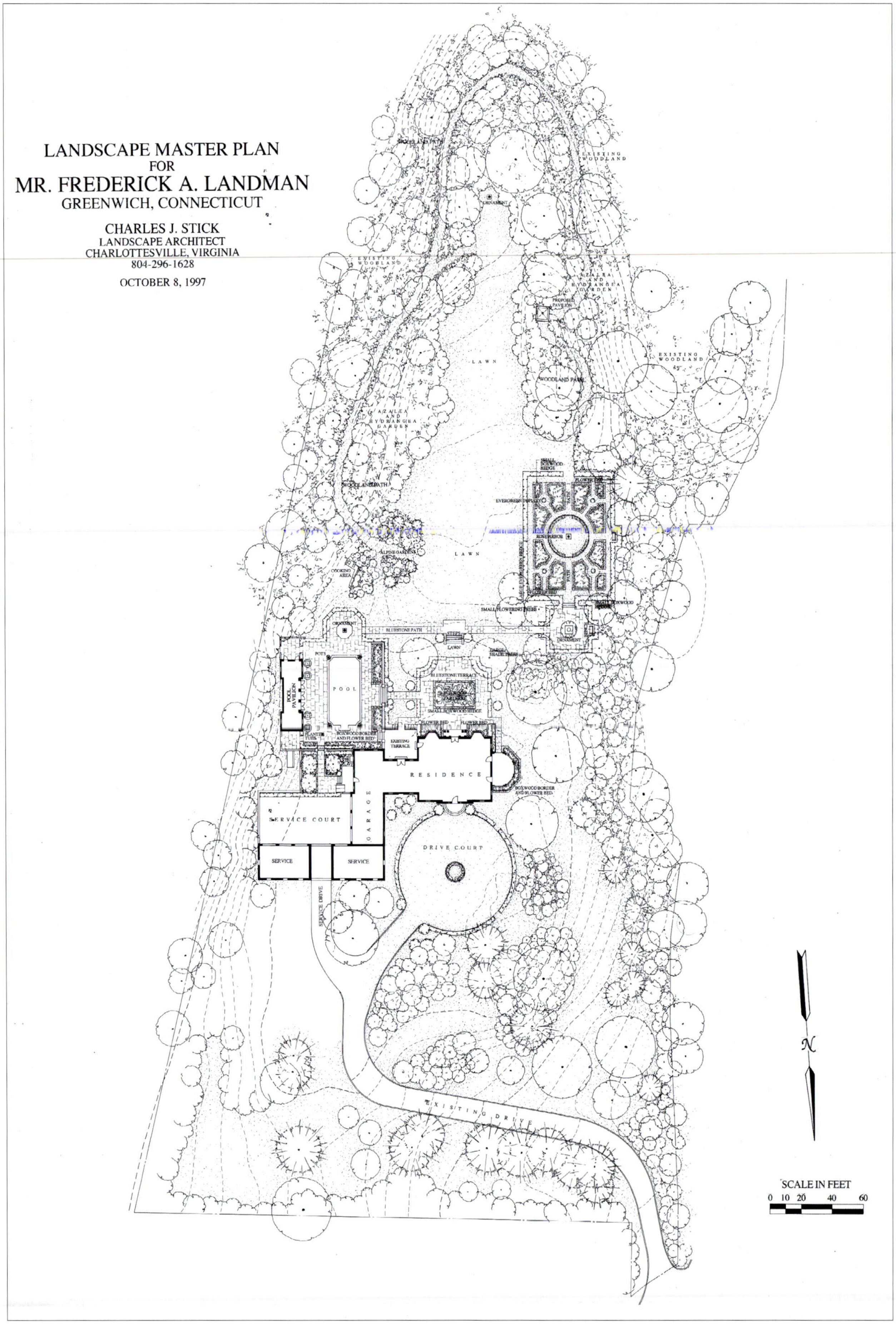
LANDSCAPE MASTER PLAN
FOR
MR. FREDERICK A. LANDMAN
GREENWICH, CONNECTICUT
CHARLES J. STICK
LANDSCAPE ARCHITECT
CHARLOTTESVILLE, VIRGINIA
804-296-1628
OCTOBER 8, 1997
LAWN
LAWN
POOL
POOL PAVILION
RESIDENCE
SERVICE COURT
GARAGE
SERVICE
SERVICE
DRIVE COURT
SERVICE DRIVE
EXISTING DRIVE
N
SCALE IN FEET
0 10 20 40 60

Left: The south terrace features a lead statue by John Cheere and a sandstone sundial.

Below: The south terrace planting plan.

Opposite: Master plan.

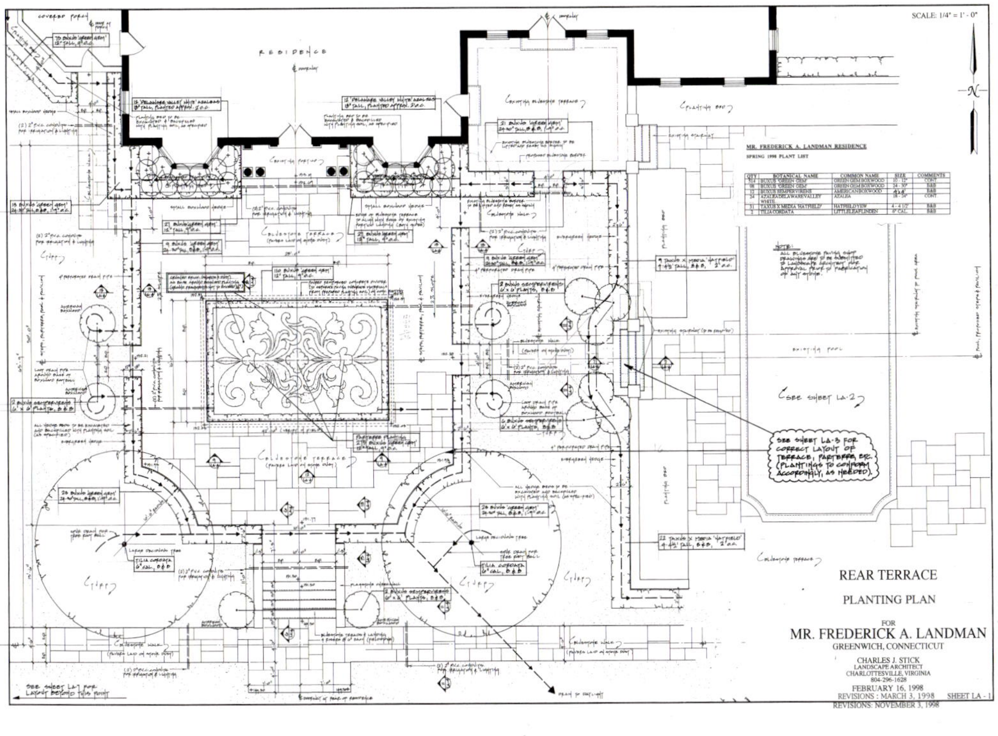

the entry to the house, removing a large oak tree that had been an impediment, and then addressed the rear where the lawn crept right up to the back door.

Stick's plan created a bluestone terrace with a boxwood parterre that transitioned the new garden room to the landscape beyond. In many ways, the terrace is the foyer for the garden experience, with avenues of exploration opening in every direction. As the garden has expanded, first from the original six acres, then to eight, ten, and now thirteen, the terrace has been not just a conduit but a gateway to discovery.

A stone wall flanks the Golden Path as it descends Clapboard Ridge amid Corylopis, *Azalea* 'Delaware Valley White,' hellebores, and Carolina silverbell.

Sleepy Cat Farm is dominated by the topography of Clapboard Ridge, a heaving mass of granite that bisects the original property Landman purchased. Its crest is a vantage point to identify "the genius of the place." There was a lawn behind the house, but much of the property was overgrown and underutilized, except by deer and foxes, hawks, and herons. Stick's plan featured the ridge and surrounding woodland by creating an elegant pathway that would encircle the property and unlock its latent opportunity.

"Early on, the idea for me was to connect landscape back to the house with the ridge and a woodland garden," Stick recalls. "The topography was extremely amenable to what I wanted to do."

Following animal trails, Stick designed what would become the "Golden Path" descending the western crest of the ridge toward the rear of Landman's land. "For me, that path is like a necklace, and there are various jewels along the way," he says.

The path, composed of crushed Virginia river rock, is clear as it descends beside the ridge, gravity and allure urging a visitor onward. A stone wall, which looks as if it once distinguished this New England farm from its eighteenth-century neighbor, flanks the left edge of the path as spent daffodils and bursting bluebells rim the right. Birdsong pervades. It's almost hard to believe these plants can flourish and bloom amid the overstory of mature hardwood trees, but then the sun emerges from a leaden sky and filters through the canopy, delivering warmth and light amid the leafy oasis. The soaring beeches and oaks provide a vertical contrast to the ground-level plantings, prompting a visitor to consider the whole scene—not just the adjacent eye-catching blooms.

"The verticality and abstract quality of those straight trunks is something I've always loved in this part of the garden," Stick says.

Nearby, the sound of water compels further discovery, and one of the more striking features emerges, as Ernest Hemingway wrote, "gradually, then suddenly." A limestone grotto fronts a massive outcropping as water cascades down mossy stones, under the path, and into a fishpond on the other side of the glen. The patina of the stones and the worn water channel suggest they hark back to a much earlier era.

On Stick's recommendation, Landman visited Valsanzibio, in Italy, which features three grottoes, but the precedent for the Sleepy Cat Farm feature is William Kent's Rousham House in Oxfordshire, England.

"I thought we could do something interesting with this rock outcropping, and Kent made stone structures in several of his gardens that had a real rustic quality to them," Stick says.

Design development sketch of the grotto.

The grotto alone could be enough, and for the neighborhood frogs, it certainly seems to be, but on either side of this cultivated cave, pathways lead up and over the opening to the outcropping above. From there, a bench provides a vantage point to survey the path, the touchstone ridge, the towering trees, and the varied plantings of the understory in their own individual stages of spring.

"One of the things that I wanted to accomplish here, and in most gardens that I work on, is to feel as though you're encapsulated by your surroundings," Stick says. "When you walk through this garden, you don't feel like you're in the suburbs of New York City."

The Golden Path continues down the slope with subtle opportunities for discovery emanating into the leafy folds beyond. One of the more novel elements of Stick's design is that visitors can be both lost and found at the same time—each path somehow closing a loop or connecting an unseen feature. Those pathways extend passive invitations to revelation, orientation, and subsequently, perspective. Many lead to the highest points of the property, allowing a visitor to survey, preview, and understand the garden unfolding below.

Inspired by Kent's rill in the gardens of Rousham (the "rill of all rills," according to Stick), the Golden Path at Sleepy Cat Farm ambles, meanders, invites, and encourages. Much as time definitively shapes many of the elements of this garden, its presence here is manifest in the path's infinite patience. The path is neither urgent nor indifferent. To linger, to contemplate, or to seek that secret, that jewel around the bend?

William Kent's gardens at Rousham were the inspiration for the grotto at Sleepy Cat Farm. Below are the eighteenth-century rill and octagonal cold bath.

Accompanied by a channel of water trickling down from the grotto, the path descends into a peaty bottom of ferns and moss and hostas. It would be easy to think this is an unintentional effect, but the design reconciles the natural conditions and topography with the overarching art. Sunlight filters through the canopy in what seems to be the leafiest part of the garden—all green and dark and dense. It is close here, but not forbidding. As dramatic and overwhelming as it can seem—when considered all at once—it is equally approachable on an individual scale. Is the garden a jaw-dropping display of floral prowess, a testament to the molding and modeling of nature, or does it stimulate deeper, more philosophical questions about beauty, joy, harmony, life?

Those things can all be true, of course, and the next set of jewels along the Golden Path offers, if not answers, at least the opportunity to ask.

Emerging from the woods into an open sky, an elevated, cedar-plank Spirit Walk juts at jagged angles above a carpet of Japanese irises. The indirect path is an attempt to evade evil spirits, unable to levy their cumbersome burdens amid such a darting course. A dense bamboo hedge at the edge of the wetland further hints at the Eastern influences to come. Farther along the path, a bamboo walk turns east, marked by a statue of the Buddha and a group of Indian prayer wheels installed

Below the ridge, the Spirit Walk zigzags through a garden filled with Japanese iris under a canopy of oaks and maples.

after a trip Landman took to the foothills of the Himalayas. The three engraved copper wheels hang on a Stick-designed scaffold, its brilliant red paint commanding attention amid the palette of natural, earthen colors.

Prayer wheels from Ladakh, India, hang on Chinese Chippendale kiosks amid hostas, viburnum, and ostrich ferns.

Initially, the Golden Path encompassed two-thirds of the property, linking the house with the farthest reaches of the land before Landman was able to acquire adjacent parcels. Even so, Stick maintains, the Golden Path is the "big idea" of the garden, and every addition and ensuing iteration has either amplified, utilized, or honored it.

"There's nothing about walking through the garden that isn't about the 'big idea' of the path," Stick notes. "There is a hierarchy to the pathways. Each offers another chance for discovery, but you're not losing sight of the main idea."

With approximately one-third of the garden unfolding to the east, the Golden Path ascends the rocky ridge, ever binding diverse garden experiences to one another, equally fulfilling its functional and philosophical duties of discovery. At this point,

Above: The Chinese Pavilion, fabricated by custom millworkers Gaston & Wyatt of Charlottesville, Virginia.

Opposite: A stone archway with a capstone "Porta Paradiso" (Gate of Paradise) marks the entrance to the perennial border garden. Fritillaria, tulips, and roses bloom in early spring.

a bronze statue of Narcissus overlooks a small basin encircled with moss-covered stones. His own vanity aside, Narcissus is looking the wrong way here. The best views are in nearly every other direction. A switchback path up the rocky ledge finds a seemingly secret door to the far end of the rear lawn, while just below, beyond a replica of a monumental ancient Roman gate, showy tulips jostle for sunlight and attention along the latest addition to the garden: an exuberant traditional perennial border.

Above the splashy color below, the Golden Path flows beneath a canopy of wisteria, emerging at an Italian bronze statue of Mercury—the Roman god of multiple constituencies—financial gain, communication, divination, and, fittingly, travelers. Beyond his raised arm, a series of stone steps climb toward the pool pavilion, the house, and rear terrace, where the garden journey could be ending or merely beginning.

PORTA
PARADISO

Granite steps, flanked by climbing hydrangea, conical Green Giant arborvitae, and massings of azaleas, reconcile the substantial difference in elevation between the main house and the eastern garden additions.

EXPANSION

Perhaps a week after the statue of Mercury had been erected—effectively "completing" the initial garden design—Landman essentially said, "That was fun. What's next?"

"Next" would play out over the coming years as Landman was able to acquire adjacent parcels of land, each presenting new challenges and opportunities for garden experiences. "I know what goes into making a garden," he explains. "I enjoy the discussion about what we're going to do. I love the implementation—the realization of what we've been talking about."

To effectively incorporate the expansion—a pair of two-acre tracts followed by a three-and-one-half-acre parcel—Stick developed a design hierarchy to organize and structure new garden experiences to complement the original.

"I think the real genius, and I credit Charles with this, is connecting all this together," Landman explains. "How do you walk through here and not think 'It's disjointed?' It's just not."

The new land was to the east and below the house and the Golden Path. From the edge of the pool terrace, Stick designed a flight of granite steps that continued the axis paralleling the rear terrace of the house. At regular intervals, sculptures of inscrutable lions and mischievous cats accompany the visitor along, but at the far end, an obelisk encircled in a boxwood hedge is the beacon that ultimately solicits descent and promises the effort will be worthwhile. From atop this stairway, a broad swath of lush grass clearly bisects this pathway at the midpoint, providing a visual resting point and a bookmark for future discovery. While the Golden Path introduces the garden experience more gradually, these sixty-five stone steps, flanked with vertical evergreens, offer not only a more direct avenue to discovery, but also essential access to the next and subsequent chapters of this garden story.

If the axis and stairs are the preface, the story of the next chapter truly begins at that grassy green bookmark.

When the first new property was acquired, it included an uninspiring kidney-shaped pool that could either be reimagined or removed. Drawing inspiration from international travels and experience, Stick envisioned the pool as round, which led Landman to consider the Rabbit Island of Valsanzibio in Italy. As the design discussion evolved, Stick found the philosophical and architectural fiber to weave the new and existing garden together.

Stick proposed a Chinese pavilion, similar to one he had seen during a recent visit to Cliveden, near London. Rather than a moated rabbit retreat, the new garden feature would invoke the chinoiserie and complement the vocabulary of the house and its Georgian traditions. An eighteenth-century affinity for Chinese motifs and ornamentation inspired designers of the era, connecting Stick's idea historically and artistically.

Conceptual sketch of the serpentine hedge between the statue of Atlas and the Chinese pavilion.

Above: A rendering of the Chinese pavilion and surrounding koi pond.

Above and top right: For inspiration, Stick and Landman considered the pavilions at Biddulph Grange in Staffordshire and Cliveden House in Berkshire, England.

Opposite: A French limestone statue of Atlas, sculpted by Andrian Melka, anchors the far end of the serpentine hedge.

Topped with a golden pineapple, the vivid red Chinese pavilion, or Ting, positively thrums with energy from across the garden, but it sits peacefully on its small stone island, shaded by magnolia and weeping cherry trees. In a grove of Japanese maples nearby, a small stream cascades at gentle intervals, trickling out among lotus and irises and feeding the koi- and lily-filled basin below.

To reach the pavilion from the garden passage, Stick designed a double serpentine hornbeam hedge that gently pushes as the pavilion pulls. At the end of the lawn, the hedge curls outward and in deference to the pavilion, successfully having delivered a visitor to its stone steps.

Sticks cites a visit to Chatsworth in Devonshire for the inspiration of the serpentine hedge. "Visually, as the sun moves through the sky, it's much more interesting than a straight hedge. And the curves of the serpentine allowed us to plant trees that create another layer of vertical richness to this horizontal axis."

While much of the Golden Path and original garden unfold under a rolling, woodland canopy, this level part of the garden largely exists under open skies—its rooms dictated not solely by topography, but by design.

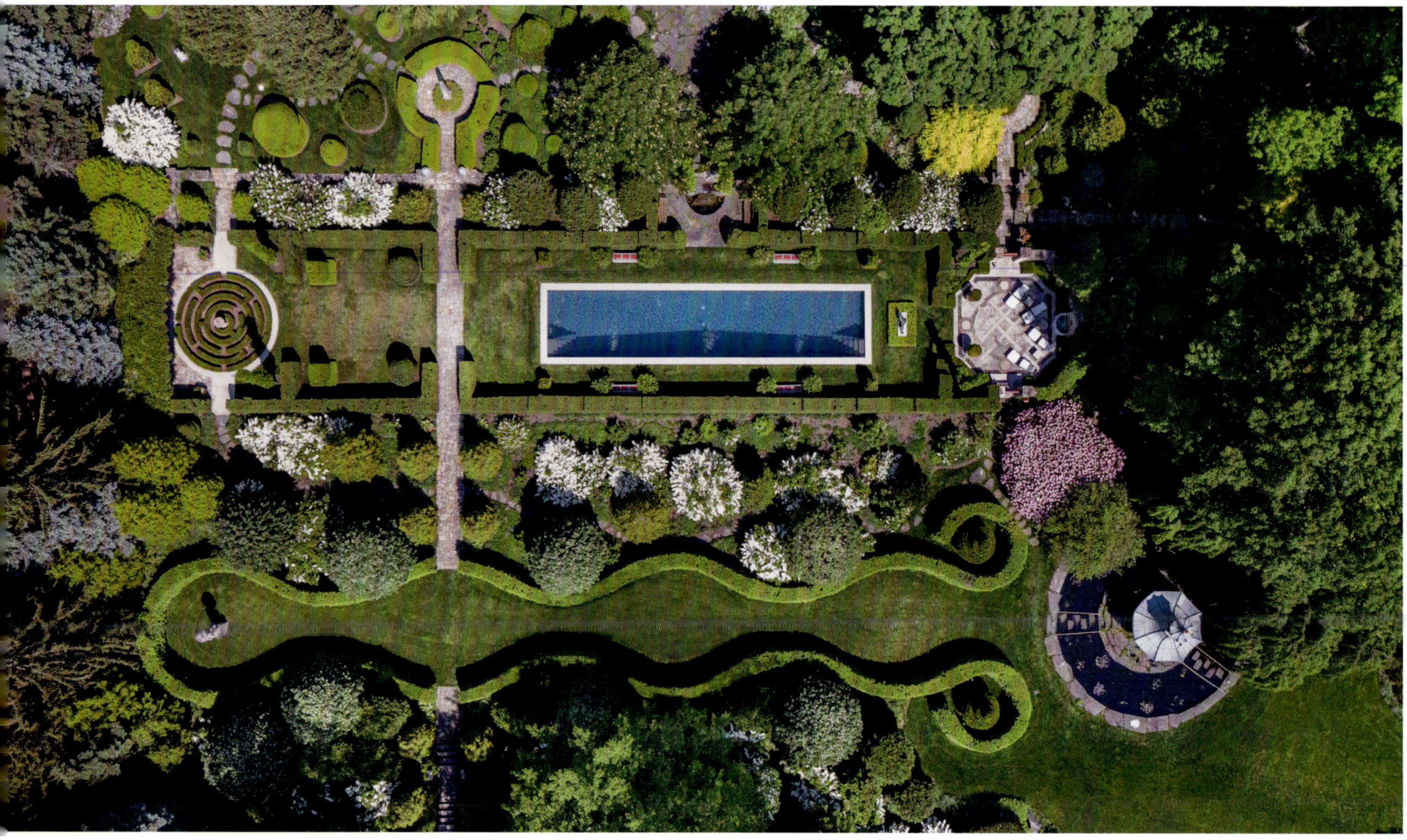

Above: From the air, the juxtaposition of parallel garden rooms and the connecting thread linking the gardens to the Barn and the main house can be clearly seen.

Opposite, above: A Chinese lantern animates the pathway between the council ring and pebble terrace.

Opposite: A terrace, designed in a pattern of limestone and beach pebbles, anchors the south end of the rectangular garden room, which terminates with the cat maze and wisteria arbor at the north end. The pool is framed by an ilex hedge with pin oaks (*Quercus palustris* 'Fastigiata') creating vertical interest.

Beyond the Chinese pavilion, a pathway divides—through a bamboo walk back toward the Golden Path or, in the opposite direction, up the hill to one of the garden's subtle high spots—the council ring.

As much as the garden is recognized for its most striking features, the connective tissue that binds it all together—the "in-between"—is equally essential and gracefully incorporated. Part of that, once again, can be attributed to the benefit of time—muting the lines of distinction between here and there—but it is also a credit to Stick's imagining of how that space might look, the secondary role it must play, and its critical function to success. The intentionally unnoticeable is remarkable in its effective anonymity. Yet a keen observer can recognize its necessity, utility, and understated elegance.

With a puff on his pipe, Stick smiles slyly at the recognition and appreciation of the in-between—the way one might when someone else notices what they, too, believe is some of their best work. From this elevated vantage point, looking down through a veil of Japanese maples, the in-between becomes more clearly understood.

"Those in-fill spaces are intentionally quite soft; that's always been important to me," Stick says. "The structure is there year-round and that's great. But those in-between spaces are the elements that add informality and vitality to the garden."

As one of Stick's favorite spots in the garden, the council ring also provides the perspective to appreciate the fundamental design concept of figure and ground. More often than not, that subject—the figure—is a solid and the area around it—the ground—is a void. When Stick and Landman finished creating the Chinese pavilion and serpentine hornbeam hedge, they began to contemplate its complement.

"How do you combine figure and ground to stimulate the eye in a positive way?" Stick asks. "The art of design is always about how those two things relate to one another. You have a figure that has an energy surrounding it and a relationship to whatever else is juxtaposed against it."

What emerged is a parallel garden room that—much like such artistic exercises—could feature the figure or ground. From literally "ground" level, this room appears to be orthogonal in contrast to its sinuous neighbor next door. But from above, the two garden spaces offer an interesting connection—each featuring rounded focal points at opposite ends with juxtaposed sculptures. Both are divided in similar proportions by the initial east-west stone path accessing this part of the garden, and each includes a secondary axis or pathway that extends the garden experience. In many ways, this parallel room is itself a pathway to the garden's next expansion.

At the south end, a pebble terrace offers an English "textural extravaganza" that provides both a place to sit and a fundamental gateway to the subsequent expansion.

When another two-acre parcel became available to the east and farther downhill, the access issue essentially had been resolved by Stick's terraced approach from the main house. However, the new site included not only existing structures, but also was bordered by Lake Avenue, posing new challenges but also offering opportunities, particularly for Lippert to add her imprint on the growing garden.

"I'm from Fresno, where you could throw out a handful of seeds and food would grow," notes Lippert, a professional chef.

This next stage of the garden's development is a tribute to Lippert's culinary craft, culminating in a small working farm, a greenhouse, and an elaborate entertaining space and guesthouse.

"When Seen arrived, she certainly appreciated the garden as beautiful, but she wondered how much of it was edible," Landman acknowledges.

With Landman and Hilton envisioning a rustic French estate as the featured structure, Stick drew on historical precedent to integrate a classic, formal garden on one side of the new building and a potager adjacent to the greenhouse and gardener's cottage on the other.

Ever mindful of the congress of the house and garden, Stick had to create one with only the notion of the other, all the while connecting it back to the original house and garden, as well as the intermediate addition.

Stick thinks about that challenge for a moment and suggests the answer to such a complex question culminates in a common theme: transparency.

The two-acre expansion ultimately revolves around the rustic, French-Norman Barn—with rough-hewn timbers, marble floor, 40-foot-high ceilings and soaring windows. Built into the side of the slope, it rises from the earth, a half-timbered, herringbone-bricked cathedral—one where Lippert can honor the breadth of fruits and vegetables cultivated and harvested outside.

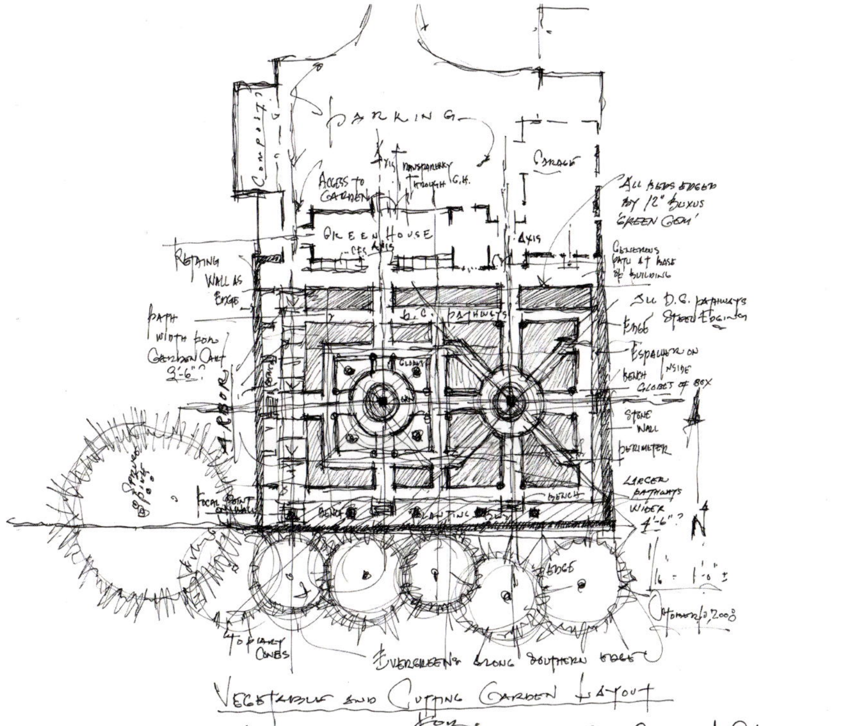

Above: To complement the Norman style of the Barn, Stick designed the landscape with crabapple trees (*Malus* 'Sugar Tyme') planted in a geometric pattern.

Left: Conceptual sketch of the greenhouse and the cutting and vegetable garden.

Above: Green Velvet boxwood parterres border planting beds in the cutting and vegetable garden; paths are decomposed granite.

To the south, an airy arts-and-crafts greenhouse provides the horticultural foundation for many of the plants—not just in the adjacent vegetable garden, but also throughout the property, with an adjacent potting shed and gardener's cottage that demonstrate the owners' commitment to the garden as a full-time enterprise.

As with most garden features, this section of the property provided both function and folly, but it represented a substantial step in Sleepy Cat Farm's evolution. "Things can be both functional and beautiful at the same time," Stick notes, citing the aesthetically pleasing but practical nature of the nearby potager.

With the addition of on-site horticulturalist Alan Gorkin, the entire garden and grounds now had a dedicated caretaker who could execute the individual and collective vision of Landman, Lippert, and Stick.

"This garden wouldn't be what it is from a horticultural standpoint without Alan," Stick says directly. "He knows plants as well as almost anyone I've ever known. Certainly, he knows plants as well as anybody I've ever worked with. And his layering of that knowledge onto this garden has made it much more interesting than it would have been on my own terms."

Above: Horticulturalist Alan Gorkin and Stick conferring in the garden.

Left: A gourd arbor constructed of western red cedar anchors the western edge of the garden.

In the potager, working from the central axis of the English greenhouse, Stick designed a geometric pattern of boxwood-lined garden beds bounded by a stone wall constructed of the same material as the Barn and adjacent outbuildings. Symmetrical plots emanate from two circular centers, each garden bed contained behind a low, meticulously clipped boxwood hedge. The effect is one of a practical parterre—particularly before the vines climb, the shoots soar, and the plants produce.

"We had a well-defined corner of the property that needed to be functional, but at the same time it could be beautiful," Stick says. "If you look at the plan, even though the greenhouse and gardener's cottage are practical structures, the garden plan is generated from those buildings—the pathways correspond to their doors. That's the transparency and connectivity."

On the garden's western flank, Stick designed a western red cedar arbor that, during the summer and fall, is enveloped with the leafy vines of overhanging gourds. On the opposite side, espaliered grapes and pears eagerly expand horizontally along the stone wall as delicate spring greens seek the sun amid the safety of their cold frame enclosures.

Above: Stick at The Deanery, which inspired the oriel window overlooking the garden.

Right: A closely clipped boxwood hedge separates the trees in front of the barn. A hornbeam hedge defines the perimeter.

Beyond the vegetable garden, Stick again invoked his experience abroad to add a block of geometrically aligned fruit trees for practical and aesthetic effect. "It was a very purposeful gesture toward the style of the architecture and my travels abroad. Every wonderful Norman country house has an apple orchard, and this is one of my favorite moments within the garden. I love the fact that it's supportive of the traditions of a Norman country house, but at the same time, I love the geometry of it. And it creates a practical buffer along Lake Avenue."

Inside the Barn is another—and perhaps the most compelling—manifestation of the transparency and connectivity Stick referenced. In the center of the Great Room, floor-to-ceiling windows offer a view west to a layered, terrace garden, with each window offering a distinct vignette and garden experience. It's purposefully composed and absent distraction, carefully linking inside and out.

"It's a magical space for me," Stick says. "I had come back from a trip in England, and I suggested to Fred that it would be great if the back of this building responded to the garden in a full-fledged way. I had just visited The Deanery, one of Edwin Lutyens's great houses. The window and the view were inspired by that trip."

"One of the key words here is 'atmosphere,'" he continues. "Since the atmosphere and style of the Barn is so different from the main house, can you still create an atmosphere that acknowledges its architecture and has a continuity with the rest of the garden?"

The answer lies in the garden outside those windows.

"When you sit in that main room of the Barn and look out, you could be in France. Even though you're not, the plants that are used, the way they are maintained and sculpted, can pay homage to the Barn but at the same time be meaningful to the rest of the garden experience."

Above: A statue of a River God is mounted on a granite pedestal outside the Barn.

Left: Rendering of the Barn and the River God and fountain.

Flanked by horse chestnuts and underlain by hostas and hellebores, the unimpeded view from the Barn's center window presents a sculpture of the River God presiding over this sanctuary of sustenance. In his familiar repose, the River God reclines atop a stone pedestal, a casual sentry at the Barn's gateway to the nearby reflecting pool.

That pool, in fact, represents one of the more practical garden applications Landman and Stick devised.

During the design phase, Stick and Landman contemplated an ornate rill or canal with fountains and water jets. Instead, practicality prevailed, and the two decided a long, narrow lap pool would be more prudent, particularly if the property surrounding the Barn ever needed to be sold. As such, the pool lends a peaceful calm to this connective passage—perpendicular to the building and parallel to the serpentine hedge and nearby Chinese pavilion. Reflecting the open sky above, the pool and its long, linear ilex hedges, funnel the eye toward the features at opposite ends—the pebble terrace to the south, and, at the opposite end, an arching wisteria arbor Landman and Stick both agree is one of their favorite parts of the garden. Under a domed canopy of wisteria, a curved Chippendale-style bench offers a quiet place to consider the discoveries not just of the garden, but nature.

Above: Aristide Maillol's *Printemps sans bras* stands among the topiary of American boxwood.

Opposite: The cat maze and wisteria arbor face the pebble terrace across the pool. The arbor design was influenced by Lutyens's arbors at Hestercombe. Stick also designed the Chippendale-style bench.

"In June, when the wisteria is in bloom, looking across the reflecting pool, you see the fountain at the end, you see the pin oaks, you see this clipped hedge, these boxwood balls ..." Landman says, his voice trailing off amid ambient birdsong and the faint but distinct sound of water moving somewhere just out of sight. "Sitting here in the shade, the fragrance of the wisteria, the bees buzzing, the hydrangea in bloom, you could sit here for hours and do nothing. At a certain time of year, that's one of the most fun things to do, and there's nothing better than this."

Appropriately, an eighteenth-century French sundial sits in the center of the dwarf-boxwood cat maze, just beyond the arbor's edge, a tangible reminder that time moves ever-so-incrementally, particularly under the sweet, shady respite of the overhanging wisteria.

THE FINAL FRONTIER?

By 2012 Landman had another opportunity to acquire an adjacent parcel and expand the garden. At the south end of the property, just beyond the far reaches of the Golden Path, a three-and-one-half-acre parcel became the latest addition to Sleepy Cat Farm. Much as the other acquisitions had, this tract would require imagination, vision, and expertise to realize a cohesive addition to the garden experience.

It also required a lot of work.

A long, narrow storage building stood at the front of the tract, while invasive trees and shrubs claimed much of the balance. Behind, beneath and beyond all that lay more of the heaving New England topography that inspired the original design while also offering the openness of the more recent additions.

"That piece of the garden was an opportunity to do things that we had not done in other parts of the garden," Stick notes. "It was a different landscape in a way, with the openness of it, while also containing the meadow defined by the envelope of the of the perimeter trees. We also had to use plants that weren't susceptible to the deer pressure."

Woodland peony, Virginia bluebells, Royal ferns, and giant snowdrops (*Leucojum* 'Snowflake') border the lower woodland walk with the Limonaia in the background.

From a new construction standpoint, Landman and Hilton were limited to the footprint of the storage shed, and devised an Italianate limonaia where citrus and other delicate plants could overwinter.

The masonry structure, with its arched windows and wrought-iron gates can seem somewhat intimidating, particularly on the north side, where it looms above one of the darkest pockets on the property. But inside and on the south side, it's entirely the opposite. Sunlight bathes Meyer lemon trees in a bright, warm corridor of tiled floors beneath a timbered ceiling. Outside, a broad, semi-circular stone terrace connects the masonry structure to the rocky New England landscape. Flanked by a subtle, but substantial, granite outcropping to the west and mounded boxwood to the east, the terrace also is a transitional space between the building's sunlit interior and the open sky and grassy meadow beyond.

As seamless as the terrace, the limonaia, and the landscape appear, it wasn't always so. The final design is a testament to the strength of the collaborative relationship Landman and Stick have developed during the past two decades. The original stone terrace had been designed and constructed—with hundreds of large bluestone slabs intricately assembled—when the two realized it simply didn't work. It was too much; it required a basin and fountain in the middle. Up came the center stones.

Rendering of the Limonaia and terrace with its central water feature.

"Fred said to me, 'That's a great idea, Charles. I wish you'd come up with it about three months earlier,'" Stick recalls. "Others with a different temperament or scope might have said they would not do that. But I think that's a metaphor for our entire relationship—the acceptance of time and that ideas evolve over time. Even though that idea did not arrive in a timely fashion, the end result was better than it would have been otherwise. That water basin completes that terrace and improves the entire setting."

A pathway from the terrace extends east toward an acre of apple, peach, plum, and apricot trees in an orchard that complements a clever, geometric "berry wedge," screened on all sides (and above) to protect the fruit from foraging birds and preserving them for Lippert's sweet treatment.

On the opposite side, a pathway winds back into the "lower woodland" where tall hardwood trees and rocky outcroppings suggest a geographical kinship with the original Golden Path but clearly enjoy their own garden identity. A damp bottom, full of ferns, fronts a jumbled, rocky ridge bounded by dwarf bamboo, redbuds, and Japanese maples. Bursting bluebells jostle for attention as they envelop a stone slab path through the woodland and past a grouping of sculpted keystones on bronze poles, stoically overlooking the understory.

Virginia bluebells and giant snowdrops line the bluestone path into the lower woodland walk.

Just as you can never get lost in this garden, you're also never alone. Classical mythological and artistic figures helpfully guard and guide—particularly in this part of the garden.

This pathway—and another on the corresponding side—convene at one of the more understated, yet duplicitous garden features. It's both significant and subtle, practical and metaphorical. At the far end of the grassy meadow extending beyond the limonaia, at the nexus of the encircling paths and, ostensibly, the farthest point of the property, Stick and Landman perfectly placed an elegant, classic garden seat. It is a nearly exact copy of an eighteenth-century William Kent design at Rousham, seamlessly integrated into this twenty-first-century space. It is both a focal point and vantage point—a jewel on the connective thread weaving throughout the garden. From the terrace outside the limonaia, the bench gently beckons across the open meadow—offering engagement and discovery. But for those having experienced the full complement of this garden and still only scratching the surface of its potential for revelation, that seat is its own sanctuary—a safe place to contemplate both this section of the garden and the entire Sleepy Cat Farm experience.

It's a rainy midweek afternoon and hundreds of visitors are expected to tour Sleepy Cat Farm in a few days. Landscape crews clip dwarf boxwood hedges in satisfying rhythm as raindrops helpfully keep time on the greenhouse glass. Gorkin, who seems perpetually armed with both an easy smile and a handful of plants, is ready.

"Why wouldn't you share this?" Landman asks rhetorically. "It's fun to watch people and see how they react. Some of them come out with their eyes glazed—like they just had a spiritual experience. I know how that feels. I love coming out of a garden that way."

More often than not, unbeknownst to many visitors, Landman is the unassuming and helpful ticket-taker at the start of the tour, suggesting routes or features to see.

"When we started making this garden, there was no thought that we would make a garden together that so many are able to enjoy," Stick says. "Because Fred's interest has become so developed and he has the resources to power it, sharing the garden has become an important aspect of his life. I think that's a testament to Fred's generosity of spirit."

Still, there's work to be done, and Landman is anxious to get back to working in the garden rather than talking about it. "Fred's level of involvement and engagement in the garden is unusual," Stick adds. "The more time you spend in the garden, the more you understand the plants—what they want to give, what conditions they need, how you can provide for them. That's part of the joy, the pursuit of gardening. It's absolutely infinite."

"Gardens will never be perfect. They're living things," Landman concludes. "It's changing. It's constant. It's an evolution."

It's time.

The garden seat, a direct reference to William Kent's benches at Rousham, is framed by fragrant Oregon grape holly (*Mahonia aquifolium*) underlain by wild geranium (*Geranium maculatum*).

WAVERLEY

SOMERSET, VIRGINIA

THE story of the Waverley garden is one of relationships—between the house, the garden, the land, its owners, and the architect.

Charles Stick sits in the garden house at Waverley surrounded by innumerable garden books, biographies, framed English garden prints, and a phalanx of tobacco pipes. Beside him, a collection of walking sticks leans against a window as the morning sun warms the room. Nearby, on a large drafting desk amid rolled-up plans and stuffed folders, boxes of seeds pledge the promise of fresh summer vegetables. On the opposite side of the room, a raft of guitars and music books suggests a hint of mystery about the man who, for the last twenty years, has called Waverley home. And just outside the door of this former farrier's cottage stands a Chinese pistache tree. It's unremarkable except to Stick, who recalls loading that sapling into the back of his station wagon and bringing it home to the farm to plant.

"There's great joy in that, thinking about that tree—now thirty feet tall. You don't get to experience that unless you're present in a place for an extended period of time."

When Stick arrived at Waverley in the mid-1990s, Charlotte Tieken and her late husband, Theodore "Teddy" Tieken, had recently purchased the venerable, but decaying, Central Virginia farm in the rural Somerset community. Once a hub for horses and hounds and social gatherings, the farm, house, and gardens seemingly laid fallow. As the Tiekens worked to reclaim the property, Stick and Charlotte Tieken discovered a gardening kinship that ultimately grew into a lifelong relationship.

Above: Charlotte Tieken with a spring bouquet of tulips, lilacs, hydrangea, and viburnum.

Opposite: An aerial view illustrates the geometric designs of the vegetable garden and the precision of the evenly spaced boxwood, along the peony walk (bottom left) and the main entry (top right).

WAVERLEY FARM

Among Central Virginia agricultural estates, the house at Waverley is somewhat understated and a bit unusual for the region—a two-story, Flemish bond brick structure built in 1847 that incorporates late Federal and Greek Revival features. By the mid-nineteenth century, Jeffersonian classicism was the prevailing architectural style in the Piedmont region, and Waverley is one of fewer than two dozen brick antebellum buildings still standing in Orange County.

On fertile farmland carved from the original eighteenth-century Octonia Grant, Waverley nobly faces the old Constitution Highway and not the Blue Ridge Mountains, which provide both the background and the frame of a picturesque Piedmont scene. Directly up the hill to the southeast, towering over the pastoral community, is the stately Somerset Plantation, constructed by Thomas Macon, husband of Sally Catlett Madison Macon. Her brother, fourth U.S. President James Madison, lived less than three miles farther down Constitution Highway at Montpelier.

In 1840 Peter Cobbs and his wife, Courtney Henshaw Cobbs, purchased 855 acres of the sprawling Somerset Plantation that would become Waverley Farm. Three years

Waverley has a rich horse and hound tradition, including hosting the closing meet of the Keswick Hunt Club, where Charlotte Tieken served as Master of Fox Hounds for seventeen years.

later, Cobbs sold 255 acres and presumably used the proceeds to build the original Waverley house. The Cobbs family sold Waverley Farm in 1863, and it changed hands multiple times before it was purchased in 1903 by James Andrews and his wife, Julia Sophie du Pont Andrews—her cousin William du Pont and his wife, Annie, had purchased Montpelier two years earlier.

Annie du Pont transformed Madison's four-acre garden of vegetables, fruit trees, flowers, and ornamental shrubs into a twentieth-century formal garden with brick walls, iron gates, statuary, and restored terraces among hearty boxwood hedges. Local lore suggests that Mrs. Andrews received boxwood clippings from the Montpelier garden to establish the framework of the Waverley garden.

When the Tiekens visited Waverley Farm in the mid-1980s looking for mid-Atlantic agricultural opportunities, they found a decaying main house at the heart of a neglected farm. They couldn't even tell that just to the west of the house was a mystical garden shrouded by paradise trees and unchecked honeysuckle.

"We drove up, and I said, 'This is it. This is it,'" Charlotte Tieken recalls.

Waverley and a pair of neighboring farms were purchased together to form a large Central Virginia agricultural holding for Babson Farms, the Tieken family agricultural enterprise. The original intent was to fix up the house and sell it as a small, fifty- or sixty-acre farm while retaining and working the balance of the agricultural acreage. But as the work on the house progressed, plans to sell were scuttled, and Waverley would become the eastern outpost of Chicago-based Babson Farms.

THE GARDEN

The lawn between the house and the garden was overgrown, as was a nearby kennel that wasn't discovered until about a year later. A multi-acre vegetable garden to the rear of the garden had been lost to nature as well. The wild Virginia lands of the original Octonia Grant were reclaiming their charter.

It was time for a new relationship between the land and its latest caretakers.

"We spent two years cleaning," Charlotte Tieken recalls. "I had no vision for the garden. It was enough to get the farm's tenant houses in order. I could not even think about the garden, other than the fact it was mysterious."

By the time farm manager Arthur Fontleroy had cleared out unwanted invasive inhabitants, the bones of the original Andrews garden emerged. "There were a few remaining flowers, and you could see the outlines of where the beds were. It left a lot to the imagination about what it must have been," she continues. "Once the Waverley houses were transformed, I became enthusiastic about the garden because you could tell it was just an amazing garden. Amazing."

Even in its raw state, the garden exuded a sense of solitude, and it was time to cultivate that relationship.

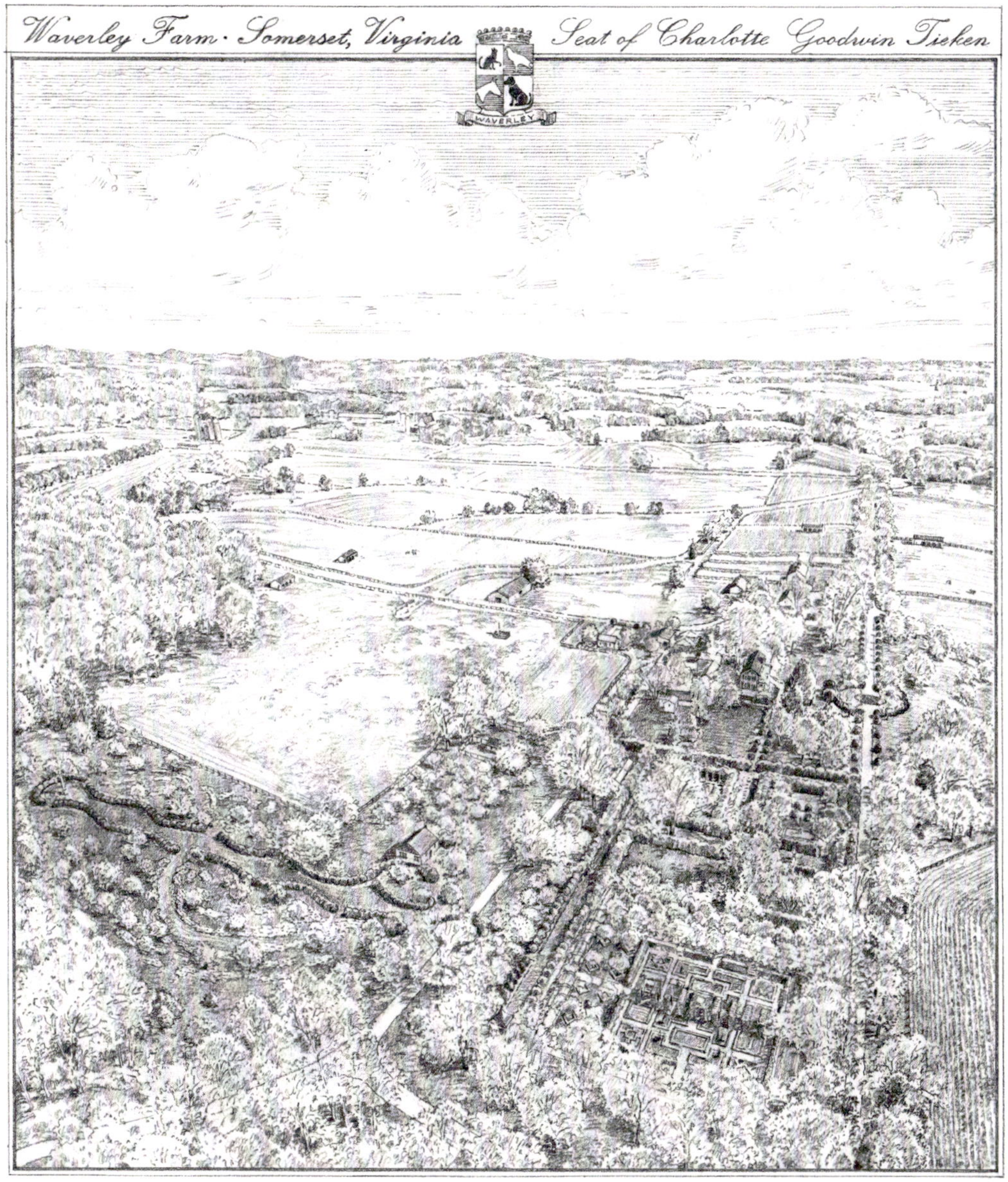

Above: Stick planted the Chinese pistache tree (*Pistacia chinensis*) that grows outside the garden house—a converted farrier's cottage. It is flanked by one of his favorite native trees, a black tupelo (*Nyssa sylvatica*).

Left: Rendering of the Waverley garden and horse farm in the manner of the Dutch artist Johannes Kip.

In the early 1990s, after the Du Ponts entrusted Montpelier to the National Trust for Historic Preservation, the presidential property hosted a talk by English garden designer Penelope Hobhouse. Following the lecture, Charlotte invited Hobhouse to visit Waverley. As the two walked through the garden, Charlotte recalls Hobhouse turning to her and saying, "This garden is magical." Together, the two worked to improve the soil and planted countless trees and flowers, forging a friendship amid the Waverley garden.

"I didn't want the sense of what was there to be lost. I wanted a continuation of the past—whatever that meant," Tieken explains. "That certainly meant not moving these big boxwood bushes and putting a lot of buildings within the garden."

By the time she was ready to elevate the Waverley garden, Tieken asked Hobhouse for a recommendation.

"I told her that I did not know how to proceed, and she replied, 'There's only one person for you to call and that's Charles Stick. He's there in Charlottesville. I've been following his work and I like what I see.'"

Above: A 1930s photograph shows the original garden rooms, whose structure is clearly visible nearly a century later.

Opposite: American boxwood globes line the new pathway connecting the house with the formal garden.

Charles Stick was wrapping up work on another Orange County estate—Mount Sharon—when he first visited Waverley. Tieken remembers leading him through the garden and feeling nervously provincial about it, pointing out overt problem areas and botanical missteps, recognizing what she didn't know, but more importantly, what she did. She acknowledged it needed help, but she was fearful of changing the spirit of the garden. Though Tieken couldn't necessarily articulate the essence of the garden, she could convey the feeling it evoked—peacefulness.

And Stick, who had said little up to that point, agreed.

Part of the mystique of the Waverley garden is its place on an active farm "village." In the Andrews era, as horse barns and adjacent kennels were built, the garden was a boxwood-ringed oasis of blooming flowers and showy shrubs. At the rear, a furrowed vegetable garden provided food for the farm and perhaps two dozen farm employees—a number of whom lived in nearby tenant houses. Even today, the Waverley garden subtly blends into the agricultural landscape. Where the garden once stood isolated, it now connects to the surrounding landscape, linking the house to the land and the farm to the house. Stick says that's part of the secret of Waverley.

"It's an overriding principle that a garden and a house are more powerful if they're a unified composition. The house becomes more important because the garden creates an atmosphere that holds the house and amplifies the spirit of the people living there out into the landscape. When I came to Waverley, there was a garden, but it had absolutely no relationship to the house whatsoever."

Together, they set about weaving the threads that would fasten those elements to one another—starting at the base of the house and working outward. One of the first steps was to link the main house to the garden. Walking up the brick pathway in front of the house, Stick looked left and saw the boxwood garden, but no clear mechanism to connect it with the house.

"There's a garden there, but how do you make it a part of the experience of being at Waverley? My first thought was a pathway that suggests you walk in that direction."

As the grade slopes from the front of the house to the garden, a series of steps

Above: A collection of magnolias, including Jane, Dr. Merrill, and Alexandrina, bloom with narcissus along the border of the serpentine walk.

Right: The bluestone terrace is surrounded by dogwoods and Doublefile viburnum, another of Stick's favorite flowering shrubs.

directs visitors toward the garden fifty yards away, but there still needed to be an engine to deliver them there. To achieve a seamless connection, Stick created the suggestion of a pathway by placing boxwoods in a rhythmically spaced manner—a botanical bridge between the house and the garden.

At the south end of the house, the doors of the breakfast room facing the garden opened directly to the grass below. Capitalizing on Waverley's interior architecture, particularly the English basement, Stick designed a terrace to create a transition from the interior of the house to the exterior, with the suggestion of discovery beyond the adjoining lawn. Creating the terraces evolved into resolving the front courtyard, where the squared-off entry area didn't connect with the adjacent elements. That solution was a softer shape, a circle of large boxwoods with an opening in the middle offering a portal to the farm fields beyond. At the same time, Stick envisioned the feeling of a canopy, particularly on the long, straight drive into the farm.

Planted with carefully spaced boxwoods—all pruned to the same height and shape—amid dozens of interspersed willow oaks, tulip poplars, and Nuttall oaks—the driveway creates momentum and arboreal rhythm, propelling the driver toward the house and garden.

"I don't think I've ever done it more gracefully than I've done it here," Stick considers. "When you come through the gate, there is a different atmosphere, an English park-like experience held together through the repetition of those boxwoods. Each creates a series of vignettes out into the field—like two vertical elements with a window in between. It's very much like walking down a corridor of a house that has windows on both sides. There is a connection to the agricultural landscape of Virginia and that, once again, is the genius of this place."

Waverley is an "inward" facing garden. Neither the house nor the garden is sited to capture the view of the Blue Ridge Mountains. Instead, the garden blends into the surrounding landscape spreading beyond its boxwood borders. To that end, cross paths bisecting the garden connect it to the house on one axis and to the land on the other.

"There are very few days I have walked through the ivy-covered piers and into the

garden that I don't feel indebted to somebody who came before me. It's a big commitment to say you're going to amplify your life by having a garden like this, and Mrs. Andrews made it part of her life's work. When I first met Charlotte, I knew it was important in her life, too."

Above: In May, peonies line the path between the boxwood hedge and the garden pavilion. Its design was influenced by Lawrence Johnston's pavilions at Hidcote in the Cotswolds.

The landscape comfortably accommodates the house and the garden. "When you walk through, it feels as though it belongs and you belong in it," Stick says. "The garden has a welcoming quality and a sense of discovery."

To amplify that sense of discovery, Tieken and Stick extended the original boxwood frame and geometry into the landscape, creating unexpected experiences. A good example is at the garden's farthest reach. At the far end of the peony walk, a small brick building beckons visitors with a familiar quality. Is it the roofline or the brickwork? Is it the open doors, offering a glimpse into the woodlands beyond? Is it all those things?

There's a good rule of thumb that as you experience a garden, you should never feel as though you arrive at a dead-end. "That was very important in how we expanded the garden from its original footprint," Stick notes.

Above: Late-spring larkspur blooms in the cutting beds of the vegetable garden with hornbeam wickets in the background.

Together, they extended the original peony walk and constructed a new pavilion with enough architectural similarity to the main house and enough transparency to make it visually engaging—even tempting. Along the path, portals lead to other discoveries within the garden and without. One of those discoveries is an arched hornbeam hedge opening to a boxwood-edged vegetable and cutting garden. In the opposite direction, the garden branches out and ascends to an altogether unexpected discovery. High upon the hill, encircled in boxwood just before the dense woodland, a towering limestone column capped by a lead urn signals the garden's peak.

From there, the scene is one of an idyllic "farm village," bisected by four-board fencing and populated with grazing horses, workshops and barns, sheds, and houses.

"Ever since I've come here, I've always had that sense of wonderment about Waverley, and I love that view up there. There's a density of built structure that has a village-like quality to it and that's kind of magic," particularly in their relationship to the main house—a theme among Virginia farms.

But how to get there? How to compel a visitor to climb the hill, to reinforce the desire for discovery without leading them to a dead-end? The answer: a serpentine experience.

Robust boxwood bushes constrict and expand up the hill, offering subtle opportunities along the way for additional exploration into flowering spring trees or budding bulbs. A towering ornament is the beacon atop this bluff that offers a pair of hidden benches upon arrival where visitors can reflect upon the scene unfolding below—the rare outward view from the inward garden.

Above: An early spring aerial view illustrates the cross-axis of the serpentine (top left), which bisects the vegetable garden (bottom right). An unseen axis continues into the adjacent agricultural field to a limestone obelisk. The garden house is at center.

Right: Millie, "the dog of Stick's life," surveys the spring blossoms from the top of the serpentine.

Above: A view from the main house to the boxwood garden where cherries and magnolias are blooming.

Opposite: Framed by a boxwood arch, a bronze statue of Mercury stands amid Iceberg roses with the main house beyond.

Most of the discoveries, though, are found within the boxwood walls of the garden rooms, where Mercury points to the heavens or Diana and an eager hound prepare to hunt. In aerial photographs, the early twentieth-century garden can be clearly seen in the modern iteration, with the original rooms of the garden neatly divided into seemingly equal parts—much of it, Tieken notes, is gratefully green—from the bright green of spring to the more muted mid-summer green. In the fall, that green tips toward silver when the lowered sun hits it just right, and by winter, it is tinged with auburn.

Even though she had not imagined having a "formal garden," it's hard to imagine a much more formal garden than Waverley, she admits. "It's got shape and space and clean lines—even with the "fuzziness" of the boxwoods. The fact they are not clipped is part of what's beautiful about it—these great green walls that change color with the seasons."

Within those walls, lush, grassed hallways extend from the stone center path, abutting the outer edges of this boxwood-ringed refuge. To the right, the figure of Diana, ever reaching into her quiver for an arrow, gazes down the peony walk through the garden pavilion, perhaps at her quarry in the woods beyond. Meanwhile, the pavilion's roof provides a useful beacon above the leafy boxwoods, orienting the visitor to position and place.

A flowering Yoshino cherry (*Prunus yedoensis*) provides a backdrop for Mercury with spring bulbs of *Anemone* 'Blue Shades' and 'White Splendor' below.

In the garden room to the left, the basin garden, koi swim amid lily pads below a statue of Narcissus, perched high enough above the water's edge to protect himself from too closely admiring his own reflection.

On the opposite side of the main walk, the winged Mercury, caduceus in hand, faces the house from within a dwarf boxwood circle. His body and upraised arm, aligned with the house, invite visitors to join him for conversation or contemplation on a nearby bench cut into the boxwood.

"That is my favorite spot," Tieken acknowledges. "It just speaks to me. There's something about leaning back on the bench, I am looking up at almost the same angle as Mercury is—pointing to the stars. It just takes me up, my mind goes up and I look to the sky. We spend so much time looking down that we forget to look up."

Nearby is a gentle fountain and parterre fronting a backdrop of golden statuary of fox and hound, cat and owl, rabbit and hawk—the farm's first tenants.

The basin garden features "The Beasties"—wooden sculptures of fox, hound, cat, owl, rabbit, and hawk mounted on Venetian gondola poles. New Dawn roses grow on swags between them, underlain with catmint (*Nepeta faassenii* 'Six Hills Giant').

In the vegetable and cutting garden, the central pathway is covered by hornbeam wickets. Green Gem boxwood hedges line the parterres, which are animated by topiary birds and Victoria Blue salvia. A rustic oak frame supports beans and peas. Purple leaf plums bloom as a backdrop.

At the next garden intersection is an Andrian Melka statue of Bacchus, alone at the end of a grassy corridor. Just beyond, past the grassy croquet pitch, pin oaks, and black gum trees, lies the geometric vegetable garden with four- and six-sided beds planted with tomatoes, squash, beans, onions, and herbs in another homage to the original Andrews enterprise. Amid the dwarf boxwood-lined beds are topiary rabbits, chickens, and squirrels.

This is Stick's favorite spot in the garden.

"It's a wonder to me that I can come home to this farm at the end of the week and have the great luxury to not drive back out of this driveway until Monday morning. Over the weekend, we eat food that has come only from this place."

Unlike other gardens of this extent and complexity, Waverley does not hew to a formal master plan. Instead, Tieken and Stick are guided by the core principles of the original Andrews garden, with subtle interventions gleaned from traveling together to gardens around the world.

"You always pick up ideas and bring them back to try. Maybe it is not the right scale or it doesn't work. Gardens are ephemeral. Some things live. Some die. Gardens develop," Charlotte says.

Whatever they've tried, it all harks back to the original composition laid out by the Andrewses in the 1920s, Stick says, suggesting that they've "refurnished" the

rooms more than redesigned them. "I think that's why the garden works."

It also works because the garden's current stewards are ever present, tinkering, planning, and planting. There is no substitute for that. They walk through the garden together and consider opportunities for enhancement—maybe a niche to place a piece of sculpture or another point of discovery—feeling the rhythms of the garden as it lives and breathes.

Charlotte Tieken and Charles Stick in front of the Trevi Fountain, Rome.

Considering the garden as a composition of sight and sense and sound, Charlotte speaks of summer evenings when the birds nest inside the boxwoods, bedding down for the night. As dusk advances, the boxwoods come alive with unseen activity. "There's something beautiful at the end of the day to think about those doves, cardinals, and tits coming home to this garden."

There's also beauty manifest in the cycle of nature—perhaps best illustrated within the vegetable garden.

"You have the digging in the offseason to get everything ready for spring, when you plant that little seed in the ground," she says. "Then, you have the picking it, the eating it. It's the entire cycle—the rhythm of what makes us exist. It's rather primal, but it talks to us, drives us, and motivates us."

By being present in place, by sharing a living relationship at the farm and the garden, Stick appreciates the opportunities it affords to experiment—not only with the shapes of things, but also with the very specific horticultural layer of the garden.

"I have an amazing opportunity, and I've explored it very purposefully," he says. "There's no substitute for having your hands in the soil. Your continued presence in a place allows you to accept the accretionary nature of a landscape and creates a confidence that the choices you make are more justified because you have a deep, enduring understanding of that piece of land. That allows a more spontaneous expression to come through. Being able to see those changes through time is the great joy that Waverley has given me."

"It's a huge joy for me, to think that this phase of my life is shared with somebody who is using this garden to promote an understanding of plants and plants that work in this area. And in this garden, it's joyful. It's totally joyful," Charlotte says. "This garden is so much about Charles and me together, about our relationship, about our travels, and the concepts and ideas from before we were together to becoming partners. It took on a different intensity surely, but the garden is so quintessentially important to us. It was the basis of our friendship and the basis of us becoming more than friends."

Above: A fresh coat of snow blankets the pyramids and globes within the basin garden.

Right: Snow clings to the hornbeam wickets that serve as the cross-axis between the serpentine and an obelisk in the agricultural field in the distance.

The Art & Craft
of Garden Making

HARRIS GARDEN

LAKE FOREST, ILLINOIS

Above: The herbaceous and cutting garden maintains architectural interest in winter.

Opposite: Green Velvet boxwood parterres define planting beds within the herbaceous and cutting garden.

HOWARD Van Doren Shaw designed Clinola, a 1911 Georgian revival house, for his Yale classmate and friend Thomas Donnelly. The property remained within the Donnelly family until 2009, when John and Stephanie Harris purchased it and asked Stick to help reclaim and enhance its rich garden history.

The Lake Forest Open Lands prairie adjoins the western edge of the site, offering a glorious view suggesting unending wilderness beyond. A low, fieldstone wall separates the formal from the informal—two stone piers serving as the open portal through which the two communicate. To the south of the house is a formal Italianate garden room designed by Ferruccio Vitale in 1927.

Stick's first step was integrating the substantial, brick house with its natural surroundings. To mediate that transition, Stick created a bluestone terrace across the rear of the house to connect with the garden room. Using Vitale's plans and surviving photographs, he was able to rebuild the trellis surrounding the space and honor the original design intent. Perennial plantings, including peonies, allium, and astilbe, offer contrasting color amid a palette of lush green boxwood, lawn, and yew. A statue of Diana de Versailles presides in an arched arbor at the far end.

Through a garden door to the east, a sunken lawn connects the formal garden to a more informal garden room, tucked between climbing roses and *Clematis jackmanii* surrounding a peaceful lily pond near the front of the property. Mindful of the beauty of the open lands, Stick sited a swimming pool opposite a formal cutting and vegetable garden on opposite sides of the main lawn. A white Chippendale-style gate embedded in a Green Mountain boxwood hedge marks the entrance to the pool. Across the rear lawn—on an invisible but very deliberate axis—a corresponding gate opens onto a dramatic and magical flower garden enclosed by hornbeam hedges. Inside dwarf boxwood provides verdant formality for beds of herbaceous perennials. Apple trees anchor corners of quadrants that radiate from the center, where white Iceberg roses encircle a stone wellhead planted with lavender and nepeta.

Originally, this site contained a much less formal vegetable and flower garden that Stick reorganized and structured, transitioning the collective space into the sensuous flower garden and an adjacent, more utilitarian vegetable garden—all within the same hornbeam hedging.

"My training would suggest the organizational principles that were used to structure that site originally were the same things we needed to tap into moving forward," Stick explains. "I was trying to envision what Shaw and Vitale would have done, how they would have organized the site around this grand house projecting itself into the Illinois prairie."

Top: Boxwood globes and a simple ground cover of pachysandra frame the view over the west terrace toward the prairie.

Above, left: *Allium* 'Gladiator' and peonies bloom in a view through the trellis of the restored Ferruccio Vitale garden.

Above: The entry drive and east facade of the main house.

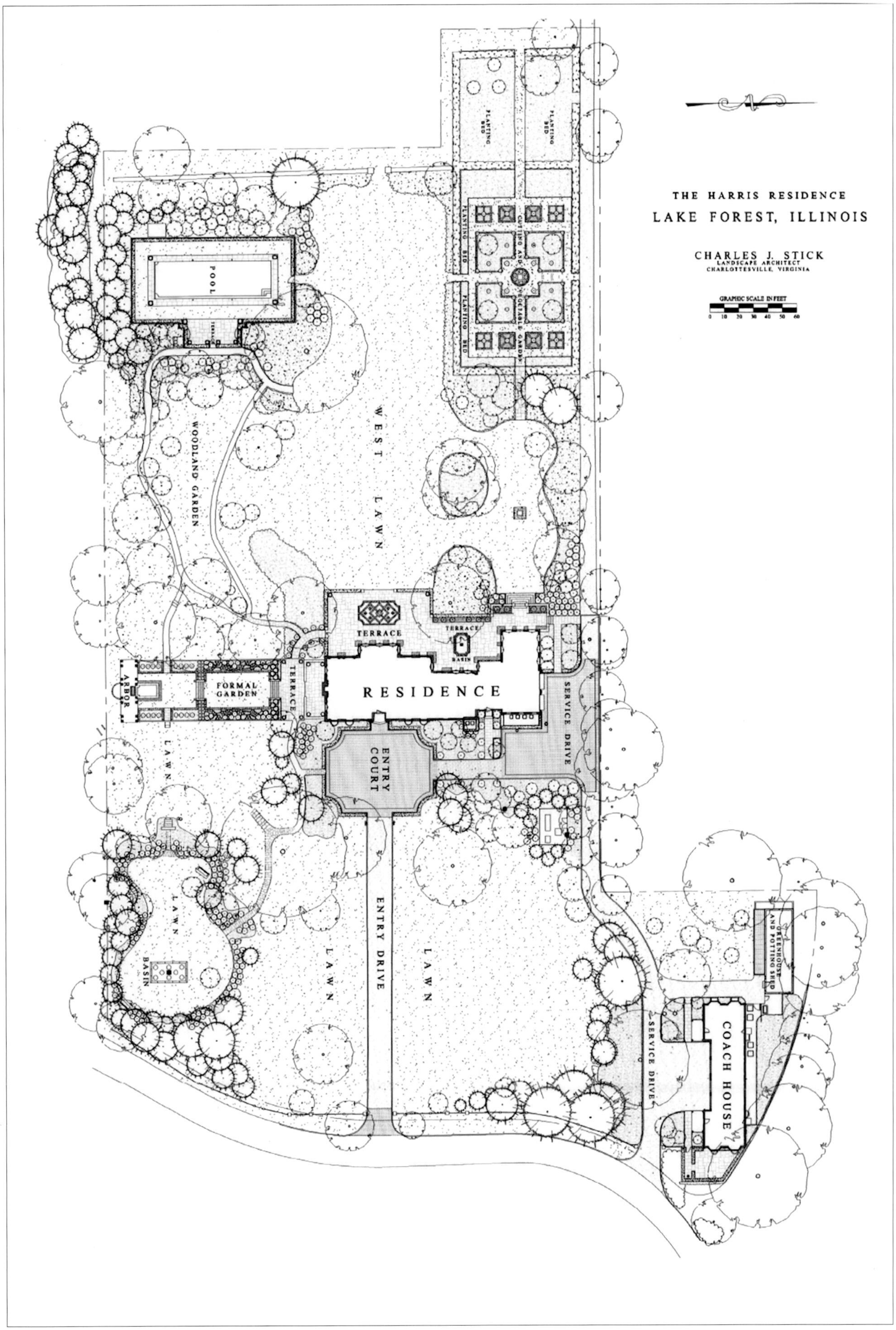

Right:
Master plan.

Above: The restored south garden, originally designed by Ferruccio Vitale in 1927.

Opposite, above: The herbaceous and cutting garden.

Opposite: The property adjoins the Lake Forest Open Lands on its western border.

Above: Oaks and maples define the west lawn.

Left: An original stone wall separates the west lawn from the open prairie.

Right: A Chippendale gate leads to the pool area amid Oak Leaf and Annabelle hydrangeas.

Below: The formality of the closely clipped boxwood parterre on the rear terrace gives way to the expanse of lawn and the open lands in the distance.

HIGHLAND PARK HOUSE

DALLAS, TEXAS

THIS garden complements an imposing limestone house designed by the distinguished English classicists Quinlan and Francis Terry that dominates the eight-acre site in the Highland Park neighborhood. "The garden needed to respond to the various facades of that powerful house, and it evoked, in my mind, a wonderful garden that Sir Geoffrey Jellicoe designed at Ditchley, a great country house in Oxfordshire."

With that precedent for inspiration, Stick created a layered, formal garden to the south of the house. The north-south cross-axis at the end of the central hall continues out into the garden, aligning it with a fountain basin end of a tapis vert lined with pleached live oaks. From the hall, the view into the garden is framed by the columns of the south portico and extends to a very distinct jet of water in a circular basin surrounded by lead urns. The double layer of Highrise live oaks provides vertical structure and privacy within the lawn—separating it from the pool terrace and nearby Preston Road. Stick admits pleached live oaks were an unusual choice and cites a Russell Page garden in Fort Worth as a precedent and inspiration.

Given the position of the house on the property, Stick had to be clever about defining the overall envelope of the property. To do so, he used existing trees, evergreen screening, and vertical layers to provide a backdrop for those elements that needed to be featured while diverting attention from those that did not.

To the east of the house, the property unfolds gently downhill toward Turtle Creek, which laps the length of the property. "The only way to treat that side of the house, in my estimation, was to have it function as the de facto English parkland that would typically surround a building of this scale," Stick says. "At the same time, we could control the view by using the existing trees to create a peaceful transition between the building and the landscape."

On the east facade, Stick created a terrace with a hollowed-out boxwood parterre, visible from the ground level, linking the greenery of the east lawn sweeping up toward the house. From that terrace, pathways diverge north and south, connecting to the service court and moving toward the greenhouse in one direction and the pool house and swimming pool in another.

To emphasize the transition between the landscape and the east facade of the house, Stick planted a pair of willow oaks and sculpted the flanking sides of both pathways as they descend to the east terrace.

On the opposite side of the house, Stick links east and west with a pathway that leads from the entry court around the outer edge of the property to the pool terrace. The path passes through a small terrace and basin garden, then meanders through trees, behind the south lawn fountain apse, and emerges at the pool

Above: The south facade of the house faces a fountain basin at the end of the lawn.

Opposite: The fountain basin is surrounded by lead urns on limestone pedestals, with pleached live oaks at the perimeter of the apse.

Above and right: The design of the south lawn is drawn from Sir Geoffrey Jellicoe's gardens of Ditchley Park in Oxfordshire, England.

Left: The east terrace overlooks an expanse of open lawn.

Below: Construction drawing defining grading and drainage specifications for the south lawn.

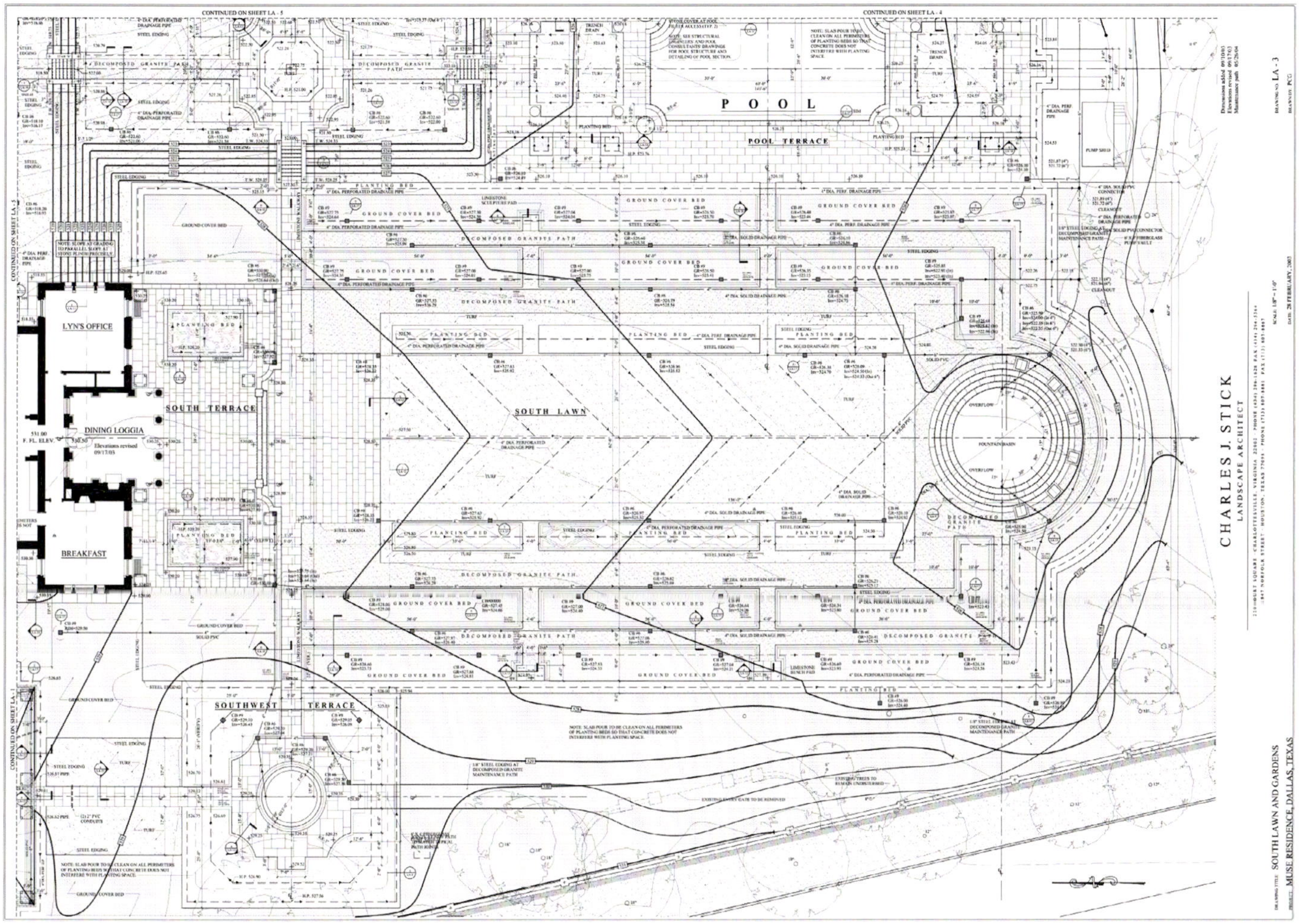

Left: Green Velvet boxwood parterres and pleached live oaks flank decomposed granite paths on either side of the south lawn.

Left, below: In spring, boxwood parterres on the south lawn are filled with annuals and narcissus.

Opposite: Master plan.

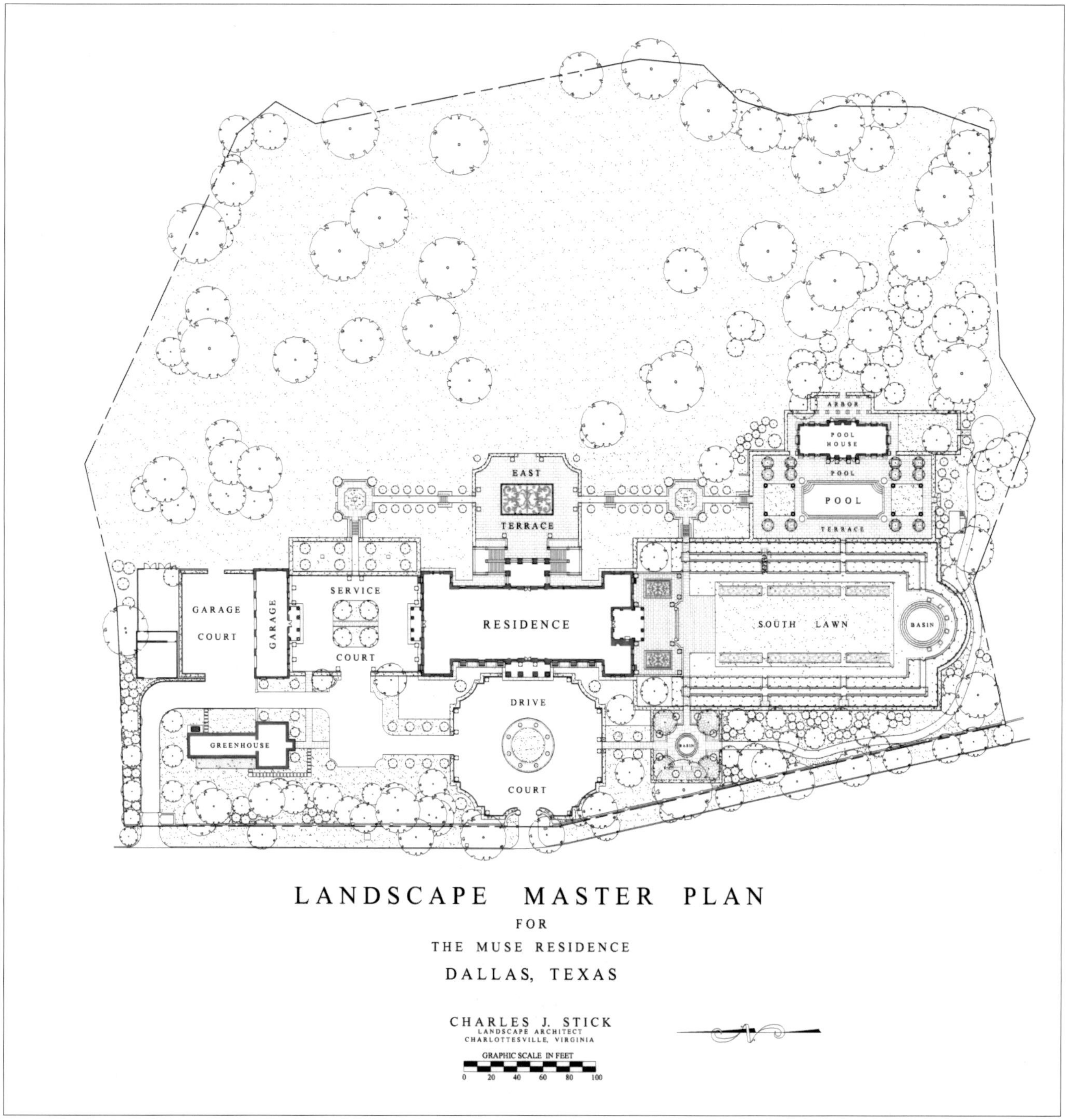

house. "That was important because I wanted the initial gesture of the south lawn to respond directly to the house, but I also needed access from the court to the pool house. This pathway through the woodland allowed that in a less formal way."

Collectively, the house, its supporting structures, and the formal garden plan fill nearly half of the property, with the balance the open, parkland setting. "This plan is necessarily simple in that the south lawn is the jewel and we wanted to stay out of the way of the house on the south side and let it be embraced by the existing, rolling landscape to the east."

THE SOUTH LAWN FOUNTAIN BASIN

AT

THE MUSE RESIDENCE

DALLAS, TEXAS

CHARLES J. STICK
LANDSCAPE ARCHITECT

Above: Aerial view of the east terrace and parterre, south lawn, fountain basin, and swimming pool.

Opposite: Rendering of the fountain basin.

WINNETKA GARDEN

WINNETKA, ILLINOIS

KING and Caryn Harris offered Stick an essentially blank slate to develop a rich garden experience around their whitewashed brick house. "I felt the nature of the house lent itself to an English cottage garden, and Caryn was very enthusiastic about that."

The ivy-walled front projects an English authenticity enhanced by the carefully sculpted boxwood globes and shrubs in the entry court. Curving beds of layered plantings provide effective depth and screening while also encouraging entry to the garden on either side of the house.

Behind a tall yew hedge the southern end of the garden is anchored by a lush rose garden. Inside low boxwood hedges, four crabapple trees rise amid an array of pink, coral, and white David Austin roses. Circular parterres with more globe-shaped boxwood and sculpted yew provide soft counterparts to the linear rose beds. The balance of the garden lies beyond an arbor covered in climbing hydrangea that acts as a functional throughway and also suggests a bit of mystery to come.

Above: A Stick-designed bench is tucked into a yew hedge in the rose garden.

Opposite: Carefully shaped boxwood parterres frame the rose garden.

"The arbor creates a shady spot to discover, but it also creates a compression of space," Stick explains. "As you emerge from the covered arbor, the garden opens again to the sky. That interplay between open and enclosed spaces is an important architectural device. That tension and release enriches the garden experience."

The arbor emerges onto a bluestone terrace, which is the exterior complement to the interior dining room. A second, offset axis continues the parallel path to the rear of the house and beyond the bay window from the dining room inside. A low box hedge mirrors the shape of the projecting window, further extending inside out where white Iceberg roses contrast with a vivid, green palette.

"The idea was to create garden rooms that respond to the house in a somewhat formal and architectural way, while at the same time developing a sense of expansiveness on a relatively small piece of property," Stick explains. "As you make your way through the house, the view out each window into the garden is a vignette. The garden rooms are very much like the rooms of the house in terms of how they are connected—by short hallways." The strategically located axes feature staggered garden rooms that suggest the idea of a larger footprint, while stimulating movement through the garden. The garden path continues through a series of circles that create an equilateral triangle, each offering access to the house or garden in various directions. Topiary peacocks point the way as peonies, delphiniums, roses, and foxgloves erupt amid the green backdrop, their colors amplified in contrast.

From the apex of the triangle, the formal garden gives way to an expansive, curving lawn between flowing beds of hydrangea and spiraea, and backed by katsura

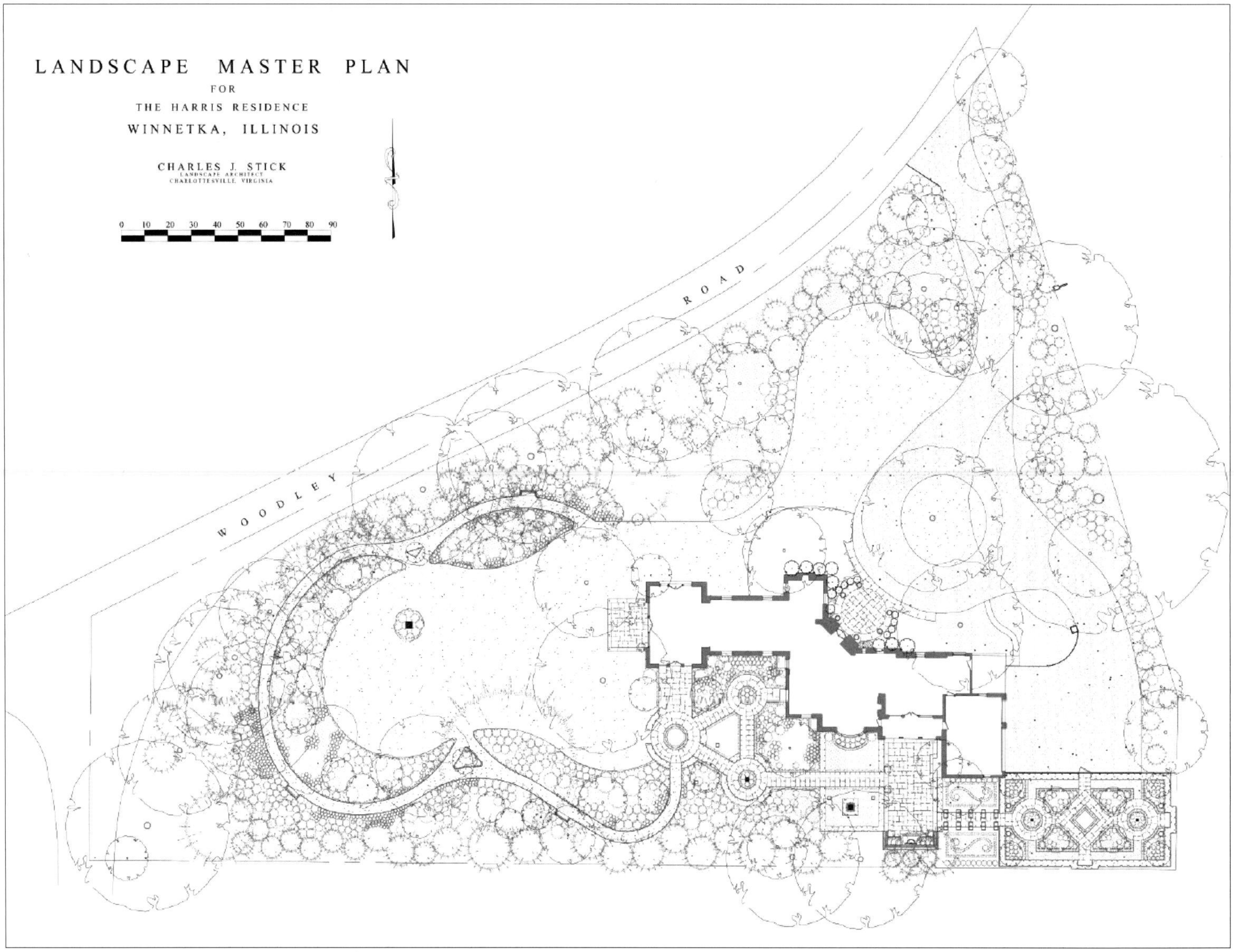

Above: Master plan.

and linden trees. Beneath the overhanging limbs of crabapples and viburnum, a decomposed granite path winds away from the house before circling the perimeter, offering glimpses back to the lawn and a covered terrace on the north side of the house.

"The peacefulness of sitting on the terrace, looking out over the lawn, is important to the overall feeling of the garden, which encompasses detailed, architectural rooms and an open lawn surrounded by curvilinear beds and the woodland path," Stick notes. "The success of the garden is that it is an example of making the most of a modestly scaled property."

Above: Beyond the rose garden, an arbor of climbing hydrangea leads to the rest of the garden with crabapple trees (*Malus* 'Snowdrift') flanking the path.

Left: Stick designed the garden rooms to correspond closely to the plan of the house so that each room offers a different vignette out into the garden.

Left and below: Herbaceous beds defined by low boxwood hedges are filled with spring and summer plantings on the south side of the house.

Stick particularly cites the work of Chicago-area landscape contractors Mike Breen and Phil Rosborough, of Rosborough Partners, who were essential to the creation of the Winnetka and Harris gardens.

Above: A decomposed granite path winds its way around the perimeter through hydrangeas and ferns and beneath horse chestnut trees.

Left: Boxwood globes define the front drive circle.

GEARY-ROSS GARDEN

SOUTHAMPTON, NEW YORK

DURING the early 1990s, Stick worked with Hilary Geary at her family's colonial revival house in Southampton. Years later, he returned to the four-acre property for a major renovation of the garden focused on two specific tasks: siting a swimming pool and a pool house and integrating two significant but very different outdoor sculptures into the landscape. Both the solutions and the overall garden experience emanate from the semicircular terrace Stick designed at the rear of the white clapboard house.

"The shape of the terrace allowed me to develop a plan that engaged the rest of the property," he explains. "From the covered porch, the curve allows sight lines from the house to radiate out into the landscape."

A central axis from the house bisects a rear lawn flanked by curvilinear hedges and terminates with a modern Li Chen sculpture as a focal point at the south end. Low, dense boxwood hedges provide the frame to celebrate the massings of hydrangea. Crepe myrtle within the sculptural apse adds mid- to late-summer color.

"Because the lawn is sizeable, I restricted the variety of plants so the garden is more about form than horticultural richness," Stick explains. "The hedges themselves are intermittently sculptural—taking the eye through the landscape, letting it rest on a plane, and then taking off again. The layering of the hedge creates the overall room for that piece of the garden—down to the fullness and color of the hydrangeas and the apse of the sculpture at the end. It's visually active and gracefully responds to the house."

The new swimming pool and pool house are sited in the southwest corner of the property near the tennis court. To reach them, Stick designed a curvilinear pathway flanked with widely spaced boxwood from the west side of the terrace. The arc of the box planting becomes the setting for three François-Xavier Lalanne sculpted sheep, which now stand in a vinca-filled crescent.

"I wanted to group them in the crook of that curve as if they were a family," Stick says. "The groundcover holds them together and, with the quizzical looks on their faces, they seem somewhat taken aback by whoever is walking by."

Beyond these two specific elements, the project extended to enhancing the parkland character of the landscape. With the help of respected nurseryman and friend Mike Graham, Stick selected 14- and 16-inch-caliper hornbeam, linden, and beech specimens from Halka Nurseries to transform the entrance and perimeter of the property. He also replaced the direct approach to the house with a curvilinear drive.

"By letting the driveway wander slightly, the front courtyard sets the stage for the presentation of this lovely house," Stick explains. "It's also a way of making the property feel larger than it is and, with the surrounding boxwood and privet, allows the garden to be a surprise. You don't perceive it until you're on the terrace behind the house and it all expands before you in a peaceful way."

Above: A Li Chen sculpture sits in the reflecting pool of the south lawn. Surrounding beds are filled with masses of Endless Summer Blue hydrangeas.

Opposite: Evenly spaced boxwood globes create a pathway from the main house terrace to the pool house. The path engages a grouping of François-Xavier Lalanne sheep.

Above: Sweeping plinths of boxwood hedges are accentuated by massings of hydrangeas.

Left: Stick designed the semicircular terrace in collaboration with noted interior decorator Mario Buatta.

Opposite: Master plan.

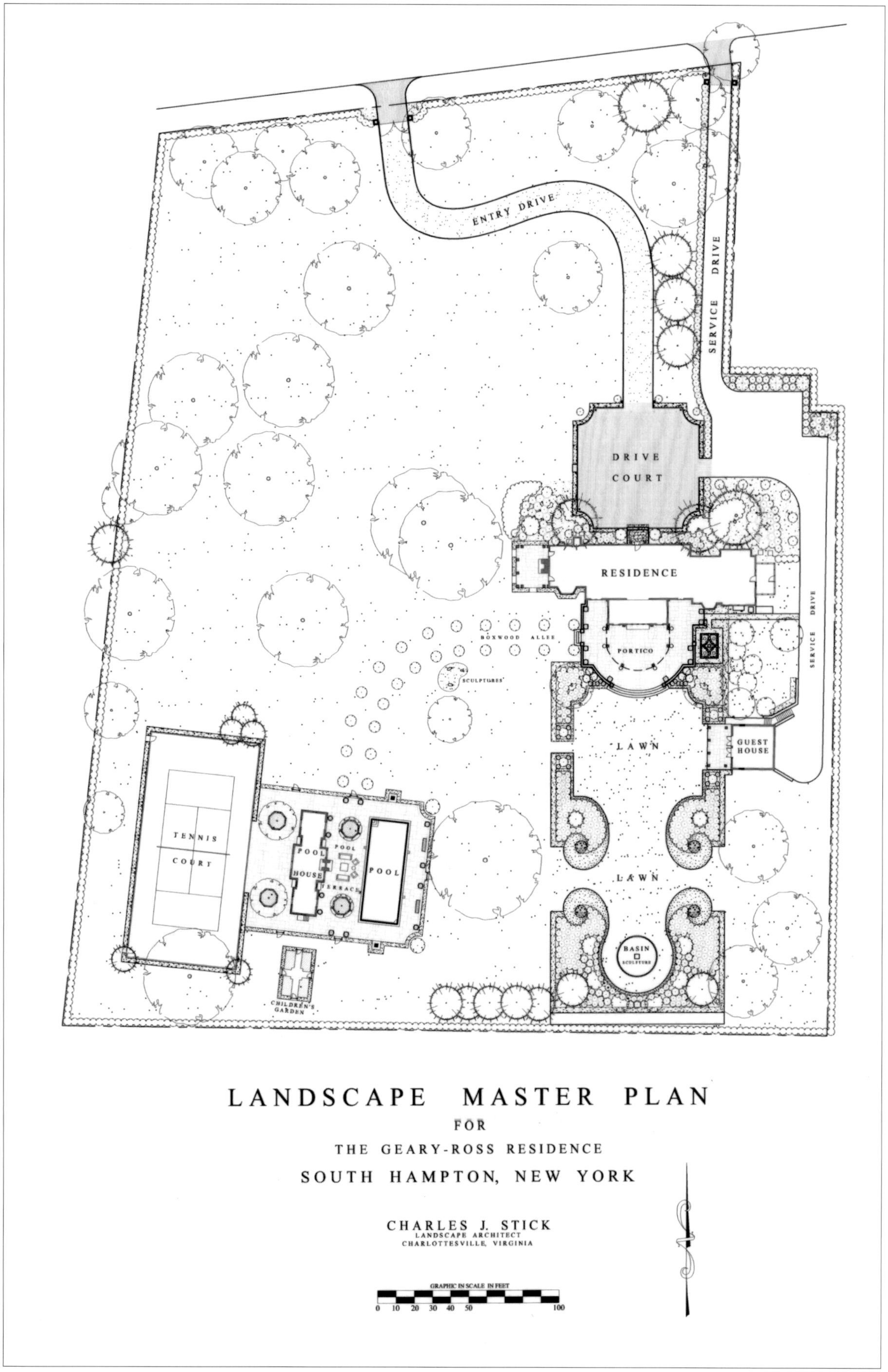
ENTRY DRIVE
SERVICE DRIVE
DRIVE COURT
RESIDENCE
BOXWOOD ALLEE
SCULPTURES
PORTICO
SERVICE DRIVE
LAWN
GUEST HOUSE
TENNIS COURT
POOL HOUSE
POOL TERRACE
POOL
LAWN
BASIN
SCULPTURE
CHILDREN'S GARDEN
LANDSCAPE MASTER PLAN
FOR
THE GEARY-ROSS RESIDENCE
SOUTH HAMPTON, NEW YORK
CHARLES J. STICK
LANDSCAPE ARCHITECT
CHARLOTTESVILLE, VIRGINIA
GRAPHIC IN SCALE IN FEET
0 10 20 30 40 50 100

Above: The sculpture apse is defined by a single row of crepe myrtles in front of a privet hedge.

Right: A plywood mock-up stood in for the sculpture during construction.

Above: A yew hedge (*Taxus x media* 'Hatfieldii') borders the south lawn.

Right: Lead urns signal the start of the pathway through the parkland setting planted with littleleaf lindens and European hornbeam.

Opposite, above: The pool terrace, with the house in the distance.

Opposite: Pool pavilion designed by architect Charles Muse.

MASSIE GARDEN

RICHMOND, VIRGINIA

THIS is one of the few city gardens Charles Stick has designed and one elevated by Will Massie's artistic intuition and design expertise.

In 1991 Massie and his sister, Annie, founded McKinnon and Harris, which has become a premier international outdoor furniture company. Stick remembers meeting the siblings for the first time when they arrived at François Goffinet's Charlottesville office with a single aluminum chair—the first in a line that launched their enduring enterprise.

Subsequently, Massie asked Stick to design a garden for his house on Monument Avenue in Richmond, a tree-lined historic street that bisects Virginia's capital city. Stick designed a small, private garden on the adjacent lot as an outdoor extension of the house, in many ways creating an external mirror of the internal plan. With existing magnolia trees buffering the property line, Stick planted an evergreen border around a lawn featuring a lead urn atop a brick pedestal, a focal point that connects the house to the garden. On a perpendicular, secondary axis parallel to the back of the house, Stick created another focal point—a fountain basin within a small flower garden. A brick pathway, edged with boxwood framing and hydrangea, connects the garden rooms on either side of a bluestone terrace enclosed by a lattice wall framed by pilasters and a molded cornice that relates to the classical detailing of the house.

"The lattice provides an airy, full-stop to this garden room, giving a sense of both transparency and enclosure," Stick says. "I think this trellised room is one of the most successful I've ever designed."

Equally, Stick credits Massie as an inspired collaborator in completing the effect. "Will understood why the architectural framework was important and amplified it with his favorite plantings and the furniture he designed," he says. "This is a wonderful example of a garden that's enhanced by the owner living there and appreciating the art and craft of garden making and all that is involved."

Above: Classically detailed trellis walls border the kitchen terrace.

Opposite: The main brick walkway leads toward a fountain basin surrounded by pleached hornbeams.

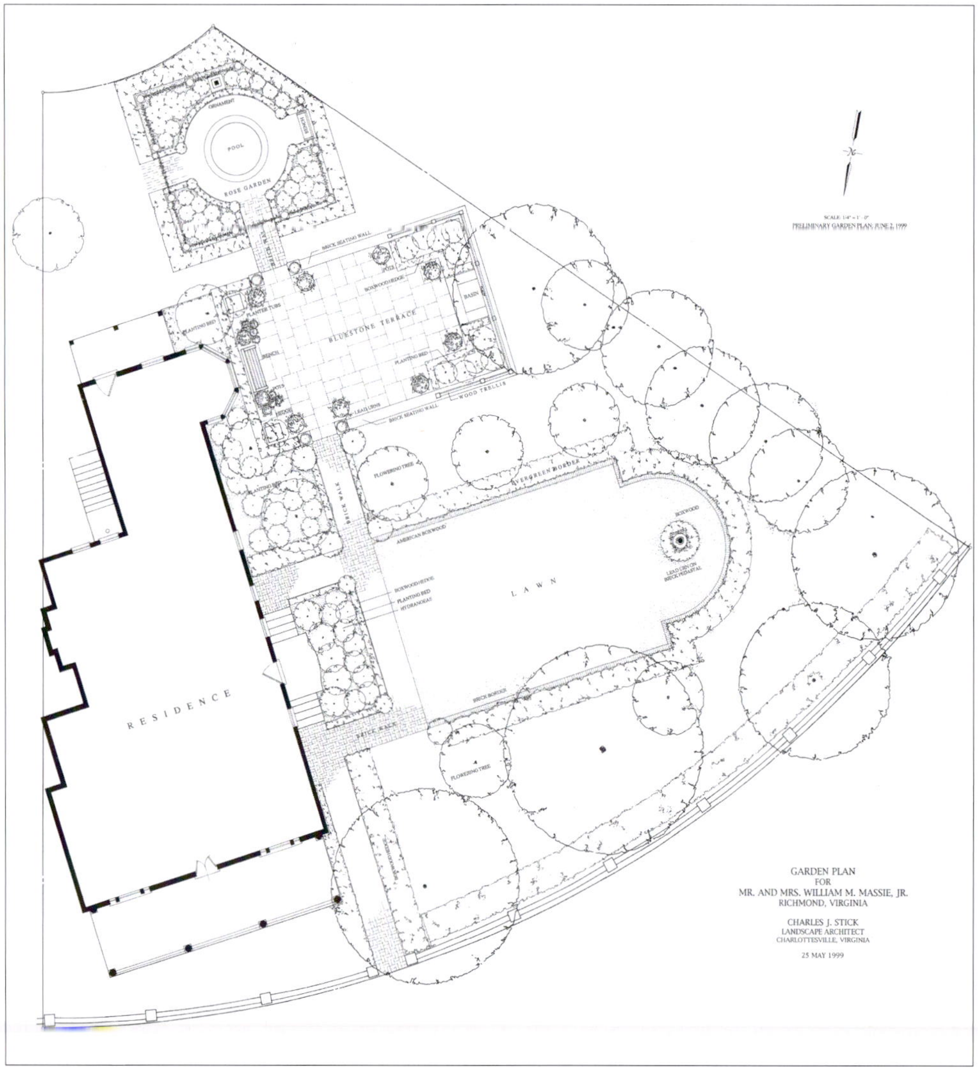

Above: Dogwoods flank the main path among closely clipped boxwood globes.

Left: Master plan.

Right: The path toward Monument Avenue with the house on the right.

Below: The kitchen terrace features Will Massie's own McKinnon and Harris furniture.

ENNISCORTHY

KEENE, VIRGINIA

ENNISCORTHY, a venerable Virginia property, was one of the early large-scale garden designs of Charles Stick's career. The Albemarle County farm dates to the late 1730s, but its red brick, Greek Revival house was constructed in 1850. When John and Robin Pickett purchased the 1,500-acre former tobacco plantation, they were committed to making it a "proper Virginia house and garden."

Stick's first task was to develop a processional entry sequence to the house. "One of the first questions I had to answer was how will guests know they have arrived," he says. "With many of these houses of this time, there is no definitive sense of arrival because of the difference between arriving by automobile rather than horse and carriage." From the main road, he designed a drive lined with evenly spaced boxwood globes that runs nearly parallel to the front of the house before reaching a circular forecourt and shifting directly toward the front door. The drive ends in a rectangular arrival court with scalloped corners and surrounded by *Ilex crenata*.

At the rear of the house, Stick installed a bluestone terrace to link the house with the surrounding landscape. Bordered by a brick wall, the terrace features planting beds of American lindens surrounded by Angelique and White Triumphator tulips. A boxwood parterre added to the side of the house is designed to be viewed from the primary bedroom suite above. Delaware Valley White azaleas, *Fritillaria persica*, and *Scilla siberica* are planted along the base of the house.

Sensitive to the history of the house and property and mindful of the view to Appleberry Ridge, Stick began incorporating and connecting the new and old elements within the landscape. To the west are the pool, pool house, and tennis court, sited to capture the view to the mountains. Stick added a winding brick pathway lined with hydrangea to connect to the house and enhanced the complex with a garden room that acts as a forecourt. The garden is planted with hydrangea, viburnum, and crepe myrtle—all of which would have been associated with a garden of the Enniscorthy era. A pathway leads to a wisteria arbor of seventeenth-century elm joists and beams in front of the tennis court. An orchard of peach and apple trees planted in the classic French quincunx pattern just beyond completes the landscape.

To the east and closer to the house lay the vestiges of an "old garden" that included mature tulip poplars and oaks and the remnants of English and American boxwood groupings. Stick's charge was to reclaim its charm and connect it to the main house. At the same time, the Picketts agreed that an old Virginia garden like Enniscorthy also should include a rose garden.

To create the new garden, Stick relied on the past, referring to *Historic Gardens of Virginia*, an illustrated compilation published by the James River Garden Club in the early 1920s. Within its pages, Stick found a plan of a garden even older than

Above: A pathway from the south facade descends toward the pool and tennis pavilion among masses of narcissus and summer-blooming hydrangeas.

Opposite: Rendering of the new entry sequence.

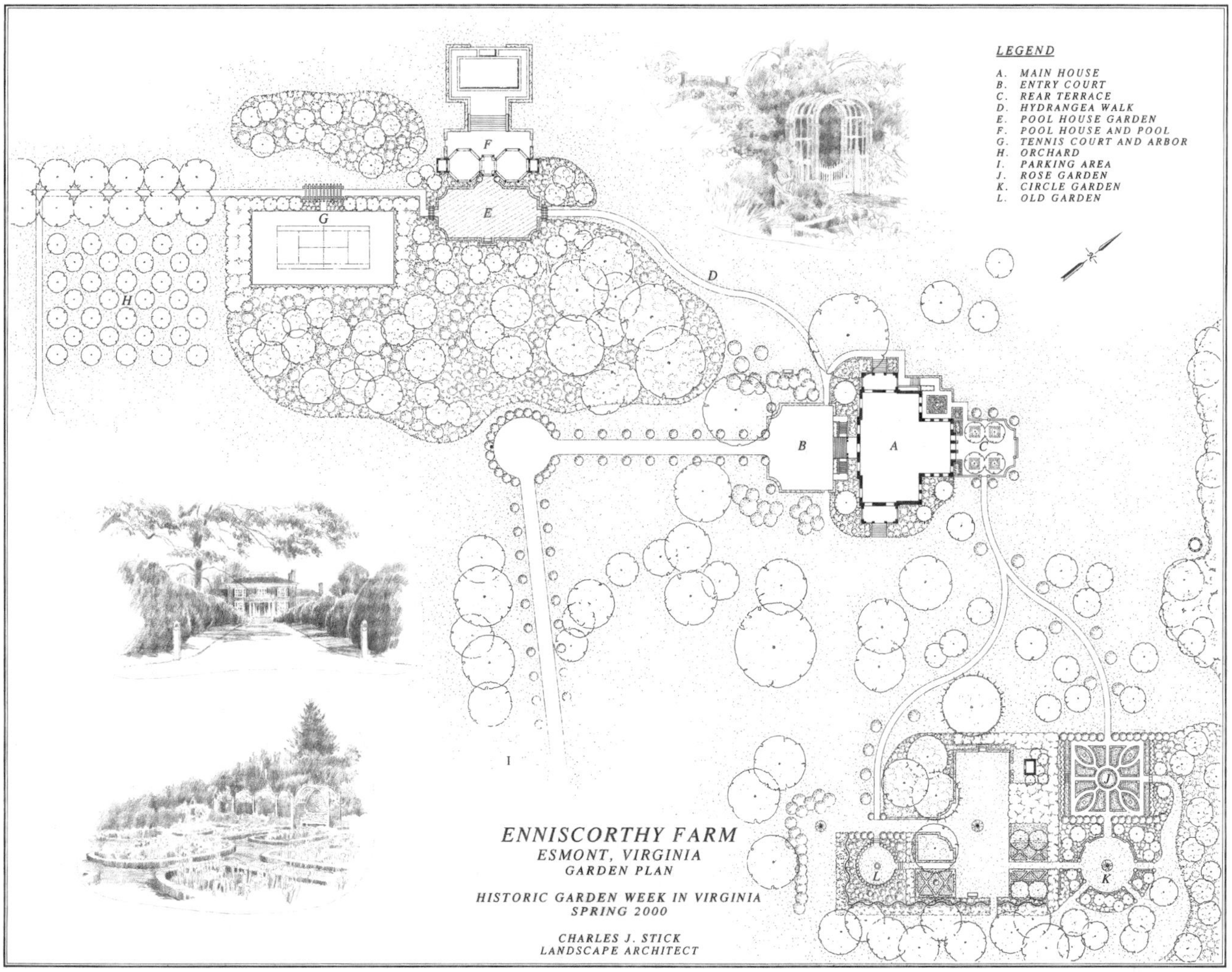

Enniscorthy—Blandfield—near the Rappahannock River in eastern Virginia.

"The pattern that emerged was flowery and wonderful and gives you a sense of a 'proper' old Virginia garden. That pattern also was portrayed in the benches we had built," Stick says. "We planted David Austin roses, and in the springtime, the beds were filled with yellow, pink, and white tulips."

Clematis and climbing roses grow in combination among tuteurs within the rose garden. On the north side is an extensive bed of pink and white peonies, while New Dawn rose swags drape from a rose bower extending outward from the garden bench and arbor.

The rose garden transitions to the "old garden" through a circle edged with beds bordered by Green Gem boxwood and filled with azaleas, hydrangeas, redbuds, and Kousa dogwoods. A boxwood arch opens into the "old garden," with boxwood parterres and classic mid-nineteenth-century Virginia garden elements, including traditional English and American boxwood, azalea, dogwood, and viburnum. Graceful, curving pathways connect the three gardens back to the terrace at the rear of the house.

Above: A garden plan given to visitors during Historic Garden Week 2000.

Opposite, above: To create a clear entry to the house, Stick redirected the driveway to a rond point so visitors could approach the house from the front.

Opposite: View west to Appleberry Ridge.

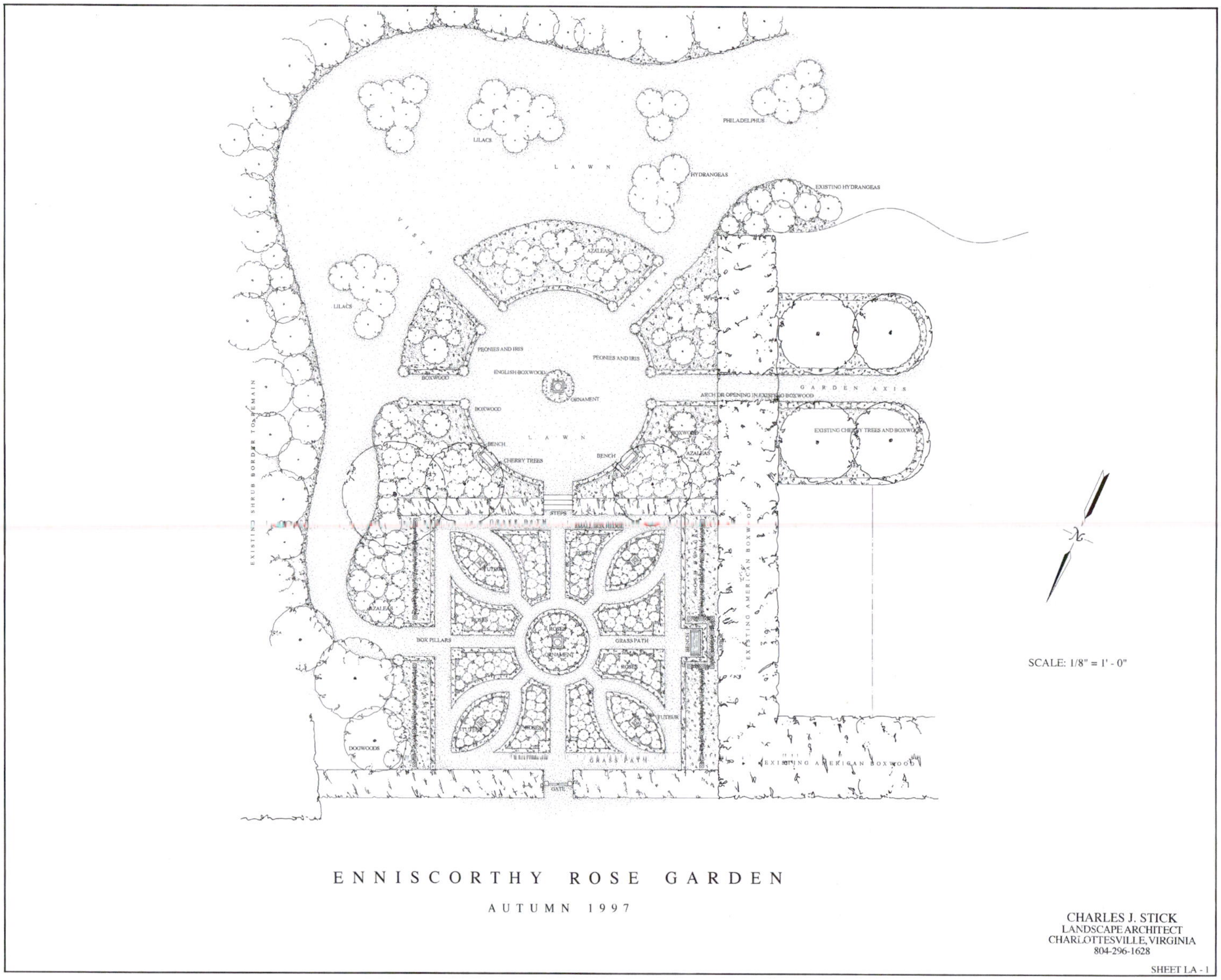

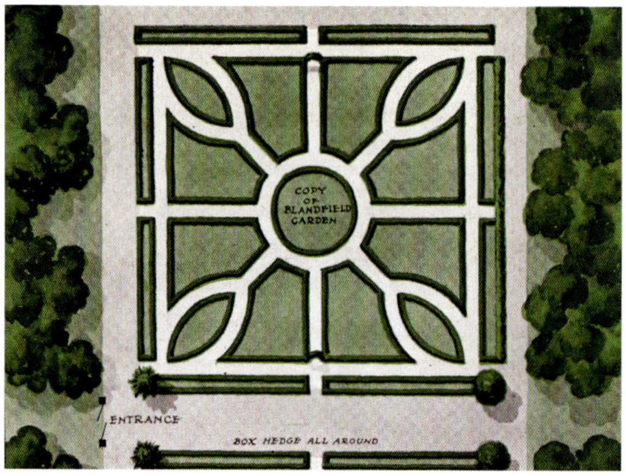

Above: Rose garden plan.

Left: The Blandfield Plantation plan that inspired the Enniscorthy design.

Opposite, above: The circular garden is planted with tulips, muscari, azaleas and hydrangeas, dogwoods, and redbuds.

Opposite, left: Stick in the rose garden during Historic Garden Week.

Opposite, right: Tuteurs within the rose garden are planted with *Clematis montana* and climbing roses. The beds are filled with David Austin roses.

LANGDON FARM

SHERWOOD, MARYLAND

LANGDON Farm traces its roots to a seventeenth-century land grant on a bent finger of land arching out into the Chesapeake Bay, west of historic St. Michaels. The imposing Georgian house dates to the 1760s on 175 acres on the water—Harris Creek to the east and Dun Cove to the south. The bay is less than two miles due west.

"I found it a very enchanting place with a majestic Palladian house set on an expansive lawn and with its proximity to Harris Creek," Stick recalls. "The way the light shimmers on the water in the morning is quite magical."

John and Suzanne Whitmore asked Stick to renovate the early twentieth-century gardens, likely created as an interpretation of the style of gardens that would have accompanied such a substantial house at its inception. According to the Archives of American Gardens at the Smithsonian Institution, Langdon Farm was noted for its extensive English boxwood garden planted in the 1920s and featured a rose garden adjacent to the house, cutting and vegetable gardens, and a traditional herbaceous garden.

"This is a proper, sturdy eighteenth-century house with a five-part facade—central block, hyphens, and pavilions. The additions in the 1920s paid close attention to the Palladian nature of the house and the colonial architecture in St. Michaels and Easton nearby," Stick says. "This is an example of a great house that gave me clues about how to respond to it and extend its energy out into the landscape."

Stick developed a master plan for the site, and he credits the Whitmores for their patient, systematic approach and fidelity to the plan. Not all of his ideas were embraced immediately—particularly the suggestion to remove two large pin oaks obstructing the facade of the house.

"I'm glad they ultimately listened to me. It was the right thing to do," he asserts.

Another of Stick's suggestions—on the opposite side of the house—also met with some skepticism. When he first visited, the land between the house and Harris Creek was planted in soybeans. According to Stick, a house on this scale, in its original conception, would have welcomed guests arriving by boat. "They all laughed at me when I told them I wanted to cut a pathway between the house and Harris Creek," Stick explains. But again he prevailed. "We cut a grass pathway from the rear of the house several hundred yards out to the water. Along that line, there were two old cedar trees, evenly spaced, suggesting it probably had been done before. That was affirmation, and that gesture was one of the best things I did."

There were no remnants of any original gardens on-site, but the re-imagined 1920s-era garden was clearly legible within the expansive English boxwood footprint to the north. "The garden rooms had been hollowed out over time, and it was our charge to refurnish those rooms," Stick continues. "We re-established a cutting and extensive

Above: Two large pin oaks were removed to open the view to the main facade.

Opposite: Iceberg roses bloom in an elliptical bed of the rose garden in front of a Blue Atlas Cedar (*Cedrus atlantica* 'Glauca').

Above: The riverfront facade of the eighteenth-century house faces southeast toward Harris Creek. Willow oaks and American boxwood border the lawn.

Opposite: Master plan.

vegetable garden within the walls of the old boxwood garden and added espaliered fruit trees and a locust-poled arbor—à la Monticello—that supported some wonderful scarlet runner beans and climbing Golden Showers roses."

At the far corner, Stick designed a dining pavilion that responded to the old pathway in the vegetable garden, flanked by cold frames and a boxwood hedge. The Flemish bond pavilion and its floor-to-ceiling Palladian windows offer unimpeded views in each direction. On the opposite side, Stick expanded the boxwood framework to incorporate a swimming pool and pool house, effectively screening them from the house.

"This was very early in my practice, but at Langdon Farm I learned to stay out of the way of changing a lovely old garden and to maintain the atmosphere that ultimately makes it special," he concludes.

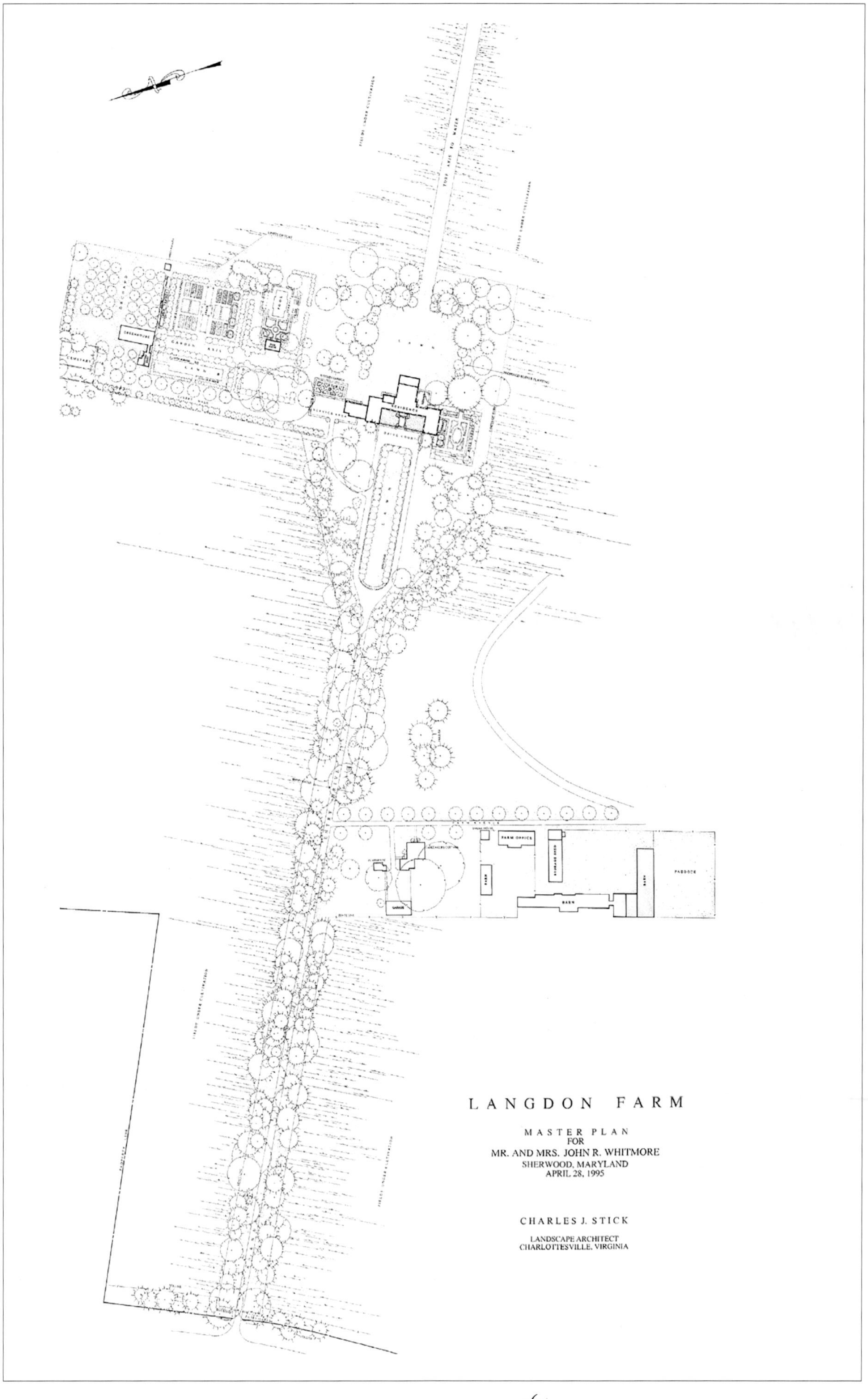
LANGDON FARM
MASTER PLAN
FOR
MR. AND MRS. JOHN R. WHITMORE
SHERWOOD, MARYLAND
APRIL 28, 1995
CHARLES J. STICK
LANDSCAPE ARCHITECT
CHARLOTTESVILLE, VIRGINIA

Above: Kwanzan cherries line the flower garden nestled within the original planting of English boxwood.

Right: A locust arbor planted with *Rosa* 'Golden Showers' in the vegetable garden.

Right: The pathway through the agricultural fields connects the house with the water.

Below: The main drive is lined with mature oaks, maples, and pines with an underplanting of spring-flowering magnolias, redbuds, and dogwoods.

MACDOUGALL-PAUSIC GARDEN

ALBEMARLE COUNTY, VIRGINIA

Above: The pathway from the main house descends through a wisteria arbor toward the pool and tennis court. Boxwood globes line the granite dust path.

Opposite: A canopy of Yoshino cherries surround the circular "node" that marks the transition from the main house to the tennis court and pool house.

WHEN Kelley MacDougall and Mike Pausic began conceptualizing their Albemarle County house and landscape in 2007, Charles Stick was there almost from the beginning.

"It was enjoyable for me because I was involved in the site planning from the start," he says. This is a great example of the way a house and a garden can work together if the process is undertaken."

On twelve mostly wooded acres west of Charlottesville, MacDougall and Pausic commissioned architect Christopher Rose to design a traditionally detailed but modern house that appears to have been expanded over time. A central brick block and gabled entrance porch anchors the front with a series of frame "additions" completing the building. To the east is a woodland dominated by pine, oak, and poplar, while panoramic views of the Blue Ridge Mountains open to the southwest.

A pair of stone piers and a gate signal the entry to the property with a driveway that graciously curves through the woodland and along the slope, offering views of the grassy meadow in the distance along the way. A dense planting of Doublefile viburnum, cascading roses, hollies, and magnolias surrounds the circular arrival courtyard. This processional sequence creates an impression of depth between the house and the road, which yields a sense of privacy.

On the far side, house, landscape, and view unfold in a similarly crafted and purposeful way.

The west facade features a semicircular covered porch bordered by a correspondingly shaped, terraced lawn. A stone wall separates the lawn from the meadow with mown grass paths arcing downhill and away from the house. "It all begins at the semicircular wall as an extension of the architecture of the house," Stick explains.

To the east, Yoshino cherry trees line a stepped, decomposed granite path descending to a circular juncture that offers a pair of options: a meadow path to the west or a wisteria arbor-lined path to the Stick-designed pool house, pool, and tennis court. The axis is definitive and creates an avenue that ties the complex back to the main house. A series of stairs absorbs the grade, while the arbor and hornbeam hedge mask the tennis court from the house.

Stick suggests the radiating pathways and the soft infill not only make perfect sense for the topography of the site but also create an armature in the plan that allows it all to recede smoothly into the landscape. "I've always been fascinated by the challenge of creating a distinct architectural gesture that could be conceived as a rigid geometrical line and tying it back into the topography of the land," he explains. "Here I did it halfway down the walkway where I needed to kink the plan because of the existing contours. A curvilinear path between these pieces of architecture would have been wrong. Instead, I designed a straight pathway with a node."

Above: A curved stone wall follows the shape of the main porch, which offers a view to the Blue Ridge Mountains to the west.

Opposite: Master plan.

This space generates a focal point that gives access to the lawn, but also directs the perspective to a long view down the wisteria arbor. The kink in the plan responds to the topography and creates an opportunity for exploring the more informal parts of the garden.

Collectively, Stick suggests this is a straightforward plan designed to integrate the structured elements naturally into the landscape. "I think that's why I really like this plan—I didn't do anything to get in the way of the view," he says. "When you're standing on the terrace, the pool, the pool house, and the tennis court are all loaded onto a section of this armature that preserves the hierarchy of the view to the distant mountains."

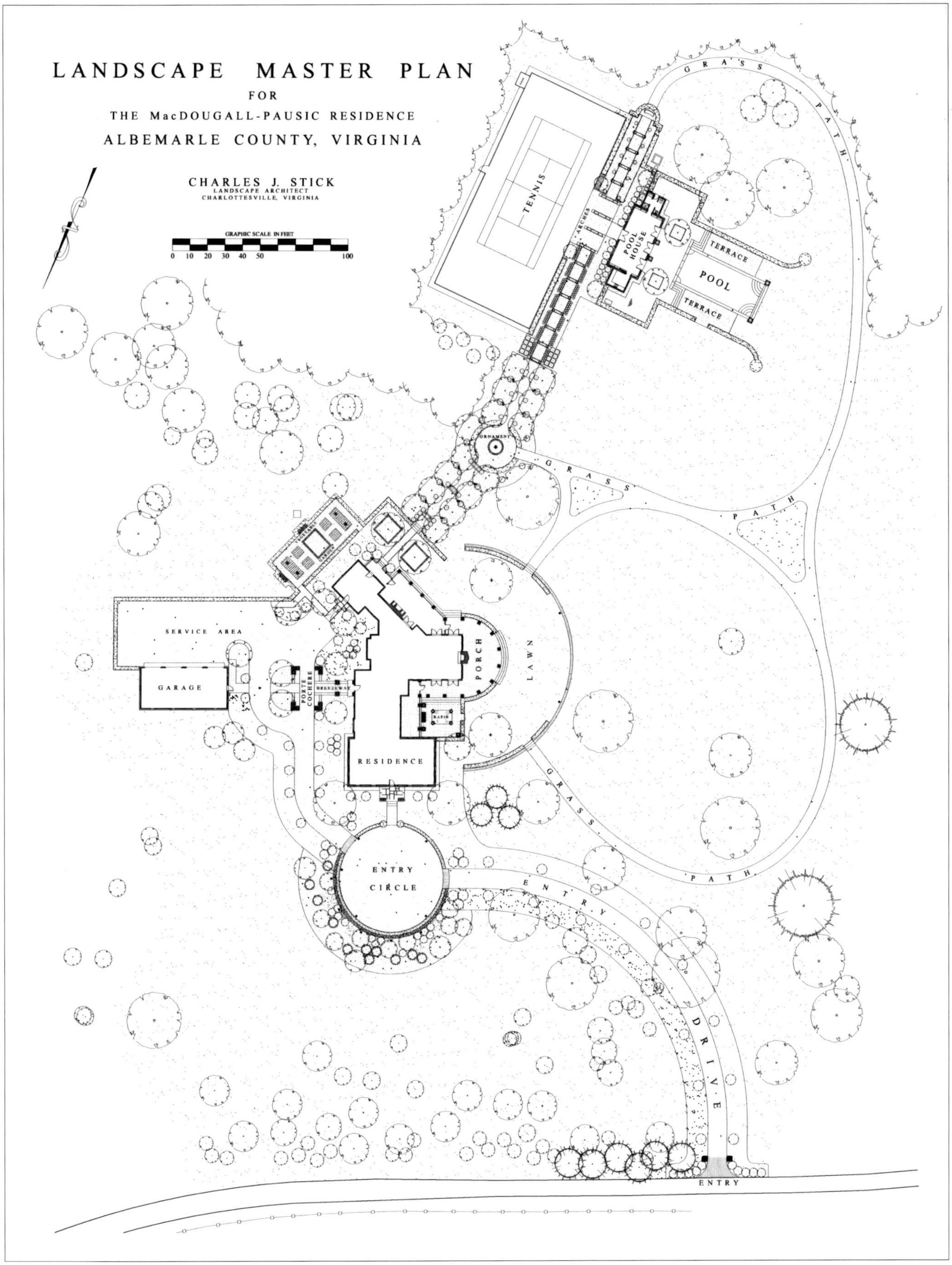
LANDSCAPE MASTER PLAN
FOR
THE MacDOUGALL-PAUSIC RESIDENCE
ALBEMARLE COUNTY, VIRGINIA
CHARLES J. STICK
LANDSCAPE ARCHITECT
CHARLOTTESVILLE, VIRGINIA
GRAPHIC SCALE IN FEET
0 10 20 30 40 50 100
TENNIS
ARCHES
POOL HOUSE
TERRACE
POOL
TERRACE
GRASS PATH
ORNAMENT
GRASS PATH
LAWN
PORCH
BASIN
SERVICE AREA
GARAGE
PORTE COCHERE
BREEZEWAY
RESIDENCE
GRASS PATH
ENTRY CIRCLE
ENTRY DRIVE
ENTRY

Above: The front drive circle is planted with Doublefile viburnum and matching willow oaks that frame the front door to the house.

Right: Yoshino cherries cover the pathway back to the main house with an underplanting of *Hydrangea quercifolia* 'Snow Queen.'

Left: A Torii arch marks the gateway to the woodland garden along a path defined by hellebores, royal ferns, and groupings of Japanese maples.

Below: The woodland garden descends through groups of Japanese maples, hydrangeas, and abelia.

ISLANDS NORTH AND SOUTH

JUPITER ISLAND, FLORIDA—ISLESBORO, MAINE

CHARLES STICK'S base in the foothills of the Piedmont in Central Virginia is nearly equidistant from two of his favorite garden-making places: Jupiter Island, Florida, and Islesboro, Maine, where he has worked with one of his longest-standing clients in a relationship that has endured more than twenty years.

As disparate as the two environments may seem geographically, topographically, and horticulturally, Stick's garden-making philosophy doesn't change. "There's no difference in my approach to anchoring a house to a garden. I'm still trying to expand the rooms of the house out into the garden," he says. "But there is also a constant reference back to the water and the horizon line."

Despite what could seem to be paradoxical projects, Stick has found commonalities in working within the two. "Nature—in its many forms—has a tremendous impact on garden making," he explains. "In Florida, there's the temporal nature of the weather, and a barrier island, by definition, moves. In Islesboro, you have harsh winters and an aggressive deer population. For different reasons, both places have a somewhat limited plant palette."

Regardless of the challenges, both places evoke specific, island-fueled emotions. "Of all the places I have worked, Islesboro is the most beautiful. There's something appealing about its rawness and the sense of isolation," Stick says. "Looking out over Penobscot Bay and back toward the Camden Hills simply lends itself to introspection."

Jupiter Island, on the Intercoastal Waterway, is more accessible, but Stick finds it no less moving. "I never tire of driving onto the island and being seduced by the graceful line of a Jamaican tall coconut palm. There's no equivalent in the north. There's simply nothing like the rustle of an ocean breeze through a coconut palm canopy."

Working with his client, Stick has undertaken similar projects in both places. They have reestablished seawalls and reconfigured entryways to allow for more aesthetically appropriate and functional approaches to the main houses. And, given the vital relationship between both houses and their respective gardens, Stick actively influenced the plans to more closely connect the form of the buildings with the landscape, always keeping the horizon line in view. In Islesboro, Stick and his client achieved that by creating a series of terraces from the hillside house down to the water's edge and adding a continuous porch across the west elevation facing Penobscot Bay. "Your perception of the horizon changes as you move vertically through the site," he notes. "You feel as though the house is a ship sailing on the water."

On the flatter plane of Jupiter Island, the connection between house and garden and the horizon line was accomplished through the plant palette. The variety of plants and the way they are grouped and massed creates interest and movement from place to place. The water always is the star of the show, so the foreground may be developed somewhat,

Above: Steps bordered by *Crinium asiaticum* climb to the garden pavilion on Jupiter Island beneath a canopy of coconut palms and pygmy date palms.

Opposite: The cutting garden at the Islesboro, Maine, cottage.

the middle ground less so, and the distance in the landscape allowed to emphasize the water and the view as the dominant feature.

The plants Stick uses in Maine are necessarily very different than the ones he uses in Florida, but the approach is the same: to achieve architectural expression in the garden. On Jupiter Island, Stick uses viburnum or pittosporum to define garden rooms. Ficus and palm trees stand in for oak and lindens with similar purpose, though to different effect. "The canopy created by a coconut palm is very different from an oak in terms of the ceiling it creates. The canopy of a grouping of coconut palms creates a more diaphanous ceiling that yields a wonderful expansiveness," he explains. "And ficus trees—which have unbelievable volume—create a sense of drama when you engage them from a distance."

Stick took a similar approach at the Islesboro property, strategically planting red oaks and maples to frame the view westward toward the Camden Hills. Other improvements included a walkway emerging from the centerline of the house and extending down the terraced lawn to the water's edge and an "old-fashioned" flower garden tended year-round by expert gardener Maxine Nelson.

"Nature expresses itself much differently in the north than in the south, but I'm still using plants in the same way to structure environments," Stick says.

Below: The west lawn overlooking the Intracoastal Waterway is defined by a ribbon of coquina stone and surrounded by a closely clipped hedge of *Viburnum suspensum*.

Above: The Maine cottage is framed with paper birch. Stairs lead down to the shore, which has been stabilized with granite boulders.

Left: The view from the main terrace across Penobscot Bay to the Camden Hills.

Above: Coconut palms shade the coquina stone–lined sunken garden with massings of ginger and Roebelenii palms in the distance.

Left: Coconut palms and Canary Island date palms surround the guest house swimming pool. Wooden tubs of citrus anchor the corners of the pool terrace.

Right: Heather and rugosa roses are planted along the water's edge amid granite stepping stones.

Below: The cutting garden.

Opposite: A grass pathway is defined by masses of rhododendron with an overstory of birch. Hydrangeas are planted at the base of the main house with an old existing vine of hardy kiwi (*Actinidia arguta*).

Above: The guest house pool pavilion and coquina stone terrace.

V HOUSE

KEENE, VIRGINIA

CHARLES Stick describes V House, in southern Albemarle County, as the most beautiful site he has worked on in Virginia.

On 3,300 acres, Jim and Bruce Murray parlayed their profound interest in early Virginia architecture into a contemporary Palladian-style villa on a mountaintop. V House honors core principles of proportion, ratio, and scale that define classical design. A central block of hand-molded brick, cut-stone, and slate, capped by a rotunda, is flanked by curved colonnades incorporating a guest house and garage.

"Jim and Bruce Murray had a very clear idea of what they wanted and where they wanted this house to sit on this land," Stick says. "They worked with extreme care to understand the land and the architecture in pursuit of something special and fulfilling."

Stick's primary challenge in integrating the building with the landscape was cultivating and unlocking discovery with a house and view of astounding proportions. From rural Fry's Path a snaking entrance drive dips into a dense woodland, curving through and into the folds of the landscape. The final curve, cut into the slope of the hill, postpones the view of the house until visitors are oriented at a fore-circle. The drive then proceeds directly between evenly spaced American boxwood toward the bright, limestone facade through an English parkland setting. The foreground is planted with native trees—oak, maple and tulip poplars—giving the house scale on its hilltop perch and preserving the pending discovery of the only-hinted-at view.

"Coming out of the woodland to that circle and then seeing the house—it's undeniable in scale and presence. But the real prize is walking through the front portico into the central room of the house, the rotunda, and through to where the space opens up to the view," Stick notes. "That unfolding delight of discovery is a combination of architecture and landscape with only the heavens beyond."

For the steep slope, Stick planned a series of terraces that practicably and artistically celebrated the expansive views to the south, west, and north. He established a radius point from the center of the rear of the building and extended it out into the landscape. The effect is a finished lawn with a seemingly infinite edge bolstered below by unseen gambioned rock walls. To anchor the house and the garden, Stick designed a cross-axis paralleling the front of the house, bisecting the entry court. To the east, the axis leads to the garage. To the west, it culminates in a garden with a semicircular apse covered by a wisteria arbor. Four parterres in simple geometric patterns are planted with hawthorn and magnolia trees. A series of steps fan out to either side, cascading down to a lower terrace punctuated by four Yoshino cherry trees in each corner. A stone border defines the edge of the square lawn that features a circular limestone basin and a elegant jet of water. On an adjacent terrace, an elliptical

Above: Matching stone pyramids terminate the entry court perimeter wall.

Opposite: The entry court is surrounded by boxwood hedging and boxwood plinths at the base of the colonnades. Stone walls in the foreground are covered with wisteria (*Wisteria* 'Amethyst Falls').

Above: The main entry drive is punctuated by American boxwood globes positioned in a parkland setting of native trees, including willow oak, black tupelo, and red maple.

Left: A construction photo taken just as the trees were planted.

Above: Master plan.

terrace offers shaded respite from the sun and an opportunity for quiet contemplation of the view.

By purposefully positioning the gardens at the end of a cross-axis in front of the house, Stick's design offered access to the view without distracting or detracting from it. "This project illustrates how the careful integration of landscape and architecture together can tell a story, create drama, delight, and ultimately, a sense of discovery," he notes.

Left: Yoshino cherries veil the view from the garden terrace to the fountain basin and the west lawn.

Below: New Dawn roses climb the stairs linking the lower terrace with the lawn on the south side of the house.

Above: The elliptical swimming pool is oriented westward.

Right: The south parterre is planted with hawthorn trees and fragrant daffodils (*Narcissus* 'Thalia')

Above and opposite: V House offers expansive and ever-changing views in multiple directions. Most dramatic are those to the Blue Ridge Mountains.

AFTERWORD

Garden making, like gardening itself, concerns the relationship of the human being to his natural surroundings.

—Russell Page

SEVERAL days each week, I drive past the lot where my family's vegetable garden once was. As with many childhood remembrances, its mythology outgrew its actual proportions. Even though our garden now exists only as someone else's back yard, it remains rich with memories of after-dinner evenings of rigorous rural industry amid the soundtrack of summer.

I was always eager to plant. My father pushed a single-wheel cultivator, furrowing a reasonably straight line from one side of the garden to the other, the earth open to receive and subsequently give. I earnestly asked how many seeds to plant and how far apart to space them, believing such symmetry and structure surely would yield success. Hardware store seeds + water + work = food.

It shouldn't be remarkable, and yet there is undeniable joy in witnessing the basic function of nature— dazzling green shoots emerging from dried husks buried in the ground—and knowing that you played a small part in creating something so simple, complex, and beautiful. Gardens are the manifestation of our individual relationships with the natural world. They are pragmatic plots of plants in places—outcomes-based, transactional operations. They are peaceful sanctuaries—public or private escapes for personal contemplation or restoration. They are delightful diversions—happy messes of trial and error, heavier on heart than horticulture. Each represents our own attempt to connect with nature and find some measure of joy in that bond.

I met Charles Stick more than a dozen years ago, soon after my wife began taking care of the horses at Waverley. I was familiar with the farm, having grown up nearby, but I was unacquainted with his work. We met formally and informally over the years, but we never had a sustained conversation, much less one about garden-making or landscape architecture.

I had recently concluded more than twenty-five years as the editor of our local newspaper when Stick and Charlotte Tieken invited us to dinner to discuss this book project. I had photographed Waverley and its gardens occasionally while my wife worked, but I had limited horticultural knowledge and even less understanding of landscape architecture. I was apprehensive, but, as a community weekly newspaper editor, I had written thousands of stories across hundreds of topics, some of which were about gardens or at least garden-adjacent. I was genuinely excited about the opportunity.

What followed was a semi-weekly crash course in landscape architecture, principles of design, horticulture, architecture, and history, with Charles Stick as my tutor. Through his library, I learned about Russell Page, Charles Platt, Lanning Roper, Beatrix Farrand, Edwin Lutyens, Gertrude Jekyll, David Adler, and Dan Kiley. I read about the distinctions between English, French, and Italian garden design and learned to identify the characteristics of each. I appreciated the concepts of art and architecture—of figure and ground, tension and release, of the hierarchy of spaces. We talked about the role of historical precedent, the essential connection between a house and a garden, and "the genius of the place" at each of his projects.

The more we discussed the art and craft of garden making, the more I began considering the landscapes around me. How do we move through the landscape? It's a question at the heart of Stick's work and now one I ask myself regularly. I may have had a sense of it in the past, a subtle perception more than an active inquiry, but now the question is imperative—an honest effort to better understand and connect with the natural world. The more I learned, the more I saw. Instead of simply observing a landscape, I was beginning to understand it, to identify those elements that promote movement, that encourage discovery, that exist purely for the sake of beauty. I began to appreciate the landscape and my connection within it much more intimately.

Each of us interacts with our natural surroundings in ways big and small every day. We move through landscapes at home, at work, in between. We connect with nature actively and passively. Whether we recognize it or not, our own bodies are much like the landscapes we navigate daily—the underlying geology of our physical selves beneath the surface topography. Are we rocky or sandy, shifting or solid, fertile or fallow? Do we grow where we are? Would we fare better transplanted? What do we need to live? What might we need to thrive? Do we complement those planted around us, supporting one another in a symbiotic relationship? Or are we competing for the same essential elements, struggling even to survive?

How do we move through the landscape—physically, metaphorically?

Charles Stick will say he is not traditionally spiritual, but in making gardens he finds a connection to something greater than ourselves. It's like holding a prism—catching a ray of light and refracting the colors as the light passes through. Each beautiful, complementary, mysterious, and accessible. He suggests that light is an offering from a higher power and bending it is a gift to share.

"Having an understanding of our place within the whole and having a pursuit which allows you to anchor yourself within that whole is a joy," he says. "Having the opportunity to bend that light in a way that hopefully connects us to the world around us is an absolute privilege."

Garden making truly is both art and craft—the combination of imagination with execution, feeling with function, aura with architecture. For more than thirty-five years, Stick has connected people with nature on a personal scale, weaving their interests and enthusiasm with his vision, training, and travels to create gardens that tell their individual stories. Through enriching the lives of others, he has expanded his own existence through shared knowledge and expertise, ideas and energy, and most importantly, genuine friendship.

In Charles Stick's office on the corner of Charlottesville's historic Court Square, there is just the right amount of disorder to illustrate structured creativity. Books are everywhere, as they are in most rooms he frequents, many chronicling the great gardens of the world he has visited or plans to. Dozens of Charles Platt etchings cover the walls. A drafting table, loaded with drawings and plans in various stages, anchors one wall, with his colleagues on the other side within easy earshot. Stick's omnipresent leather briefcase sits atop his desk, an antique table, confirming everything here is completely and comfortably classical.

He has here almost everything he needs to connect people with nature: the sum of his training, artistic eye, travels, experience, and inspiration. All that's missing is nature itself, the glorious, elusive, beautiful variable that makes the entire ephemeral garden-making exercise worthwhile.

—Jeff Poole

The basin garden at Waverley, defined by Green Mountain boxwood pyramids, Green Gem boxwood globes and parterres, and century-old boxwood hedges.

COMMISSIONS

David Adler Cultural Center
Libertyville, Illinois

Mr. and Mrs. John Allen Jr.
Clarksburg, West Virginia

Mr. and Mrs. Philip Allen
Bedminster, New Jersey

Mr. and Mrs. Thomas Allen
Richmond, Virginia

Mrs. Patricia Altschul
Charleston, South Carolina

Ms. Janice Aron
Charlottesville, Virginia

Mrs. H. W. Bailey
Jupiter Island, Florida

Mr. Johnny Baker
Independence, Texas

Mr. and Mrs. Joshua Bank
Rye, New York

Lord and Lady Conrad Black
Palm Beach, Florida
Toronto, Ontario

Mr. and Mrs. Julian Bowen-Rees
Richmond, Virginia

Mr. Norman Brenbridge
Charlottesville, Virginia

Mr. and Mrs. Michael Brooks
Jupiter Island, Florida
Lawrence, New York

Mr. and Mrs. K. Dane Brooksher
White Oak Farms
Elsberry, Missouri

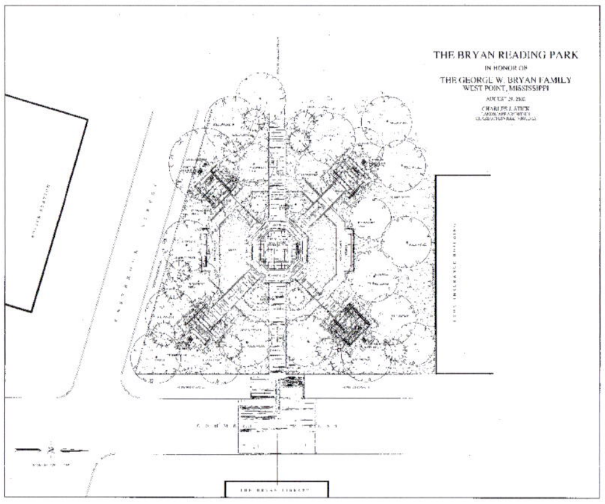

The Bryan Reading Park
West Point, Mississippi

Mr. and Mrs. George Bryan
West Point, Mississippi

Mr. and Mrs. James D. Bryan
West Point, Mississippi

Mr. and Mrs. John H. Bryan Jr.
Crab Tree Farm
Lake Bluff, Illinois

Mr. and Mrs. John H. Bryan III
Nashville, Tennessee
West Point, Mississippi

Mr. and Mrs. John Stewart Bryan
Richmond, Virginia

Ms. Julia P. Bryan
West Point, Mississippi

Dr. and Mrs. Paul Bullock
Richmond, Virginia

Mr. and Mrs. Gil Caffray
Greenwich, Connecticut

Mr. and Mrs. Robert Calhoun
Jupiter Island, Florida

Mrs. Wayne Calloway
Greenwich, Connecticut

Mr. and Mrs. Edmund Carpenter
Greenwich, Connecticut

Mr. and Mrs. Andrew Carter
Scottsville, Virginia

Mr. and Mrs. John Chadwick
Nashville, Tennessee

Mr. and Mrs. William Coleman
Tivoli
Gordonsville, Virginia

Mr. and Mrs. Charles L. Collins
Fredericksburg, Virginia

Mr. and Mrs. James Collins
Merriewood Farm
Somerset, Virginia

Mr. and Mrs. Hunter Craig
Charlottesville, Virginia

Mr. and Mrs. Archibald Craige
East Belmont Farm
Keswick, Virginia

Mr. and Mrs. Mac Crawford
Nashville, Tennessee

Mr. and Mrs. Allen Cunningham
Charlottesville, Virginia

Mr. and Mrs. Dennis T. D'Antonio
Greenwich, Connecticut

Mr. and Mrs. Alan Dabbiere
Hickory Hill
McLean, Virginia

Mr. and Mrs. Robert W. Daniel
Brandon Plantation
Spring Grove, Virginia

Mr. and Mrs. Sandy Reisky de Dubnic
Free Union, Virginia

Mr. and Mrs. Allen de Olazarra
Charlottesville, Virginia

Mr. and Mrs. Kim Des Marais
Disputanta, Virginia

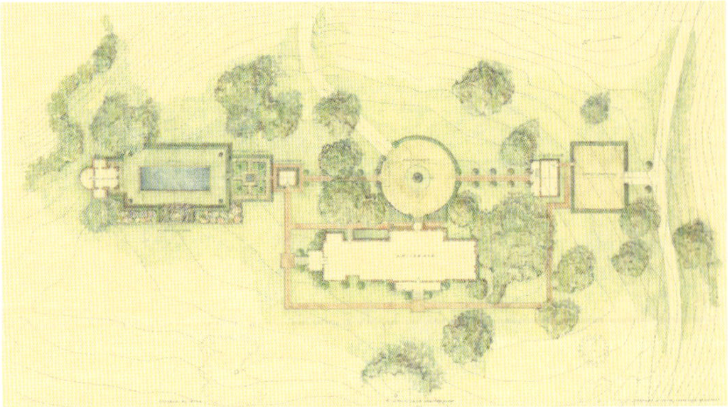

Mr. and Mrs. Rupert de Vink
Hatton Grange
Scottsville, Virginia

Mr. and Mrs. Frank Easterly
Bellona Arsenal
Midlothian, Virginia

Mr. and Mrs. Harry Easterly
Richmond, Virginia

Mr. and Mrs. Richard Egbert
Greenwich, Connecticut

Mr. and Mrs. David Emery
Nashville, Tennessee

Mr. and Mrs. Albert Ernest
Charlottesville, Virginia

Mr. and Mrs. William S. Farish
Lane's End Farm
Lexington, Kentucky

Mr. and Mrs. William S. Farish Jr.
Lexington, Kentucky

Mr. and Mrs. Temple Fennell
Charlottesville, Virginia

Mr. and Mrs. Marshall Field
Jupiter Island, Florida

Mr. and Mrs. Eugene Fife
Cherry Hill
Charlottesville, Virginia

Mr. and Mrs. Lawrence Flinn
Cricket Hill
Greenwich, Connecticut

Mrs. Patricia Gerber
Lake Forest, Illinois

Mr. and Mrs. Robert Gerry
Houston, Texas

Mr. and Mrs. Jon Gerstenfeld
Washington, D.C.

Mr. and Mrs. Christopher Getty
New York, New York

Mr. and Mrs. Charles Gilman
Charlottesville, Virginia

Mr. James Goodwin
Mill Neck, New York

Mr. and Mrs. William Goodwin
Richmond, Virginia

Mr. and Mrs. Floyd D. Gottwald
Richmond, Virginia

Mr. and Mrs. Bruce C. Gottwald Jr.
Richmond, Virginia

Mr. Boyden C. Gray
Washington, D.C.
Jupiter Island, Florida

Mrs. Gordon Gray
Jupiter Island, Florida

Mr. and Mrs. Robert Gray
Greenwich, Connecticut

Mr. and Mrs. Achille Guest
Powhatan Plantation
King George, Virginia

Mr. and Mrs. Bruce Gupton
Rocklands Farm
Gordonsville, Virginia

Mr. and Mrs. Louis Hager
Farmlands
Cooperstown, New York

Mrs. William C. Hancock
Richmond, Virginia

Mr. John Hancock
Richmond, Virginia

Mr. William Harlan
Napa Valley Reserve
St. Helena, California

Mr. and Mrs. Robert Harrell
West Point, Mississippi

Mr. and Mrs. John Harris
Lake Forest, Illinois
Chicago, Illinois
Jupiter Island, Florida

Mr. and Mrs. King Harris
Winnetka, Illinois

Mr. and Mrs. Monty Harris
Free Union, Virginia

Mr. and Mrs. James G. Harrison III
Coggan's Point, Virginia

Mrs. Sylvia Heidenberg
Englewood, New Jersey

Mr. and Mrs. Robert Heidenberg
Demarest, New Jersey

Mr. and Mrs. John A. Herdeg
Mendenhall, Pennsylvania

Mr. Jeffery Hildebrand
Houston, Texas

Ms. Lisa Hogan
Portaferry Farm
Afton, Virginia

Mr. and Mrs. J. R. Hutcheson
Richmond, Virginia

Mr. and Mrs. Charles Johnson
Windsor Hill
Charlottesville, Virginia
Richmond, Virginia

Mr. and Mrs. Michael Kane
Portsmouth, New Hampshire
Jupiter Island, Florida
Lake Forest, Illinois

Mr. Scott Kasprowicz
Willow Oaks
Middleburg, Virginia

Christina and Paul Kazilionis
Jupiter Island, Florida
Dark Harbor, Maine
Nantucket and Boston, Massachusetts

Mr. and Mrs. John Keller
Pebble Beach, California
Winnetka, Illinois

Mr. and Mrs. Alec Kempe
Richmond, Virginia

Kenmore Plantation
Mr. Vernon Edenfield, Executive Director
Fredericksburg, Virginia

Mr. and Mrs. Mark Kington
Grelen Farm
Orange, Virginia
Alexandria, Virginia

Mr. and Mrs. Alexander Klatskin
Englewood, New Jersey
Water Mill, New York

Mr. and Mrs. Charles Klatskin
Englewood, New Jersey

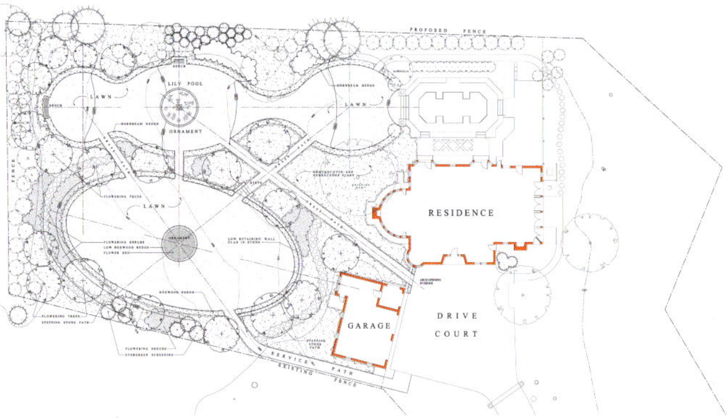

Mr. and Mrs. Henry Kleeman
Lake Forest, Illinois

Mrs. Patricia Kluge
Albemarle House
Charlottesville, Virginia

Mr. and Mrs. Lester Knight
Winnetka, Illinois
Vero Beach, Florida

Mr. and Mrs. Liam Krehbiel
Lake Bluff, Illinois

Mr. and Mrs. Frederick A. Landman
Greenwich, Connecticut

Mr. and Mrs. Clark Lee
Frederick, Maryland

Ms. Linda Leffingwell
Lake Forest, Illinois

Mr. and Mrs. John Lewis
Sherwood Farm
Charlottesville, Virginia

Mr. Todd Leap and Mr. Mark Little
Villa Maria
Lynchburg, Virginia

Mrs. Nancy H. Marcantonio
Westport, Connecticut

Mr. and Mrs. William M. Massie
Richmond, Virginia

Mr. and Mrs. Tom McCullough
Dallas, Texas

Mr. and Mrs. George McFadden
New York, New York
Southampton, New York

Dr. and Mrs. David McGroarty
Richmond, Virginia

Mr. C. Wilson McNeely
Morrowdale Farm
Charlottesville, Virginia

Mr. Henry Mendler and Ms. Suzanne Lawton
Little Merifields Farm
Keswick, Virginia

Mr. and Mrs. J. Irwin Miller
Jupiter Island, Florida

Mr. and Mrs. Jamie Minnis
Lawrence, New York

Ms. Deborah Minor
San Francisco, California

Mr. and Mrs. Halsey Minor
Fox Ridge Farm
Charlottesville, Virginia

Mr. and Mrs. Robert Monk
The Plains, Virginia

Mr. and Mrs. Richard Morrow
Skye Fall
Charlottesville, Virginia

Mr. and Mrs. James Murray
Keene, Virginia

Mr. and Mrs. John R. Muse
Highland Park House
Dallas, Texas

Mr. and Mrs. William Mynatt
Lake Forest, Illinois

New Dominion Bookshop
Mrs. Robert Troxell
Charlottesville, Virginia

Mr. Richard Nuckols
Richmond, Virginia

Mr. and Mrs. Gordon Ogden
Lawrence, New York

Old Waverly Golf Club
West Point, Mississippi

Mr. and Mrs. Thomas C. O'Neil
Lake Bluff, Illinois

Dr. Carl Outen
Richmond, Virginia

Ms. Madeline Parrish
Richmond, Virginia

Mr. and Mrs. Jerry Pate
Pensacola, Florida

Kelley MacDougall and Mike Pausic
Charlottesville, Virginia

Mr. and Mrs. John Payson
Jupiter Island, Florida

Mr. and Mrs. C. D. L. Perkins
Richmond, Virginia

Mr. and Mrs. John O. Pickett Jr.
Enniscorthy Farm
Esmont, Virginia
Southampton, New York

Mr. and Mrs. James Mathews Pope
Richmond, Virginia

Mr. and Mrs. Michael Prentiss
Charlottesville, Virginia

Mr. and Mrs. Frederick Prince
Marshall, Virginia

Mr. and Mrs. William Pritchard
The Homestead
Hot Springs, Virginia

Mr. and Mrs. C. B. Robertson
Crozier, Virginia

Mr. Gregory Robins
Richmond, Virginia

Ms. Carolyne Roehm
Weatherstone
Sharon, Connecticut
Charleston, South Carolina

Mr. and Mrs. Wilbur Ross
Washington, D.C.
Southampton, New York

Mr. and Mrs. T. D. Rutherfoord
Charlottesville, Virginia

Mr. and Mrs. W. L. Sanders Jr.
Richmond, Virginia

Mr. and Mrs. Richard Scruggs
Oxford, Mississippi

Mr. and Mrs. Charles H. Seilheimer Jr.
Mount Sharon Farm
Orange, Virginia

Mr. and Mrs. Marc Shore
Greenwich, Connecticut

Shoreacres Club
Lake Bluff, Illinois

Mr. and Mrs. Howard Simpson
Lake Forest Illinois
Harbour Island, Bahamas

Mr. and Mrs. William S. Stuart
Charlottesville, Virginia

Mr. and Mrs. John B. Syer
Charlottesville, Virginia

Mr. and Mrs. Edwin Thoet Jr.
Williamsburg, Virginia

Mr. Victor Tolkan
Washington, D.C.

Mr. and Mrs. Bronson van Wyck
Wadmalaw, South Carolina

Mr. and Mrs. Alex Vogel
The Brick House
Upperville, Virginia

Mr. and Mrs. Ted Weschler
Jumping Branch
Charlottesville, Virginia

Mr. and Mrs. Mark Wheless
Afton, Virginia
Richmond, Virginia

Mr. and Mrs. John R. Whitmore
Langdon Farm
Sherwood, Maryland

Mr. and Mrs. Julian Williams
Richmond, Virginia

Dr. and Mrs. Mason Williams
Richmond, Virginia

Mr. Eli Zabar and Ms. Devon Fredericks
New York, New York

NOTABLE BOOKS FROM CHARLES STICK'S LIBRARY

Jane Brown, *Lanning Roper and His Gardens* (Rizzoli, 1987).

A. S. G. Butler, *The Architecture of Sir Edwin Lutyens* (Country Life, 1950).

J. C. Chepherd and G. A. Jellicoe, *Italian Gardens of the Renaissance* (Ernest Benn, Ltd., 1925).

Kenneth Clark, *Looking at Pictures* (John Murray, 1960).

Michael Dirr, *Dirr's Encyclopedia of Plants & Shrubs* (Timber Press, 2011).

Franklin Hamilton Hazelhurst, *Gardens of Illusion: The Genius of Andre Le Notre* (Vanderbilt University Press, 1980).

Gertrude Jekyll, *Home and Garden* (Longmans, Green & Co., 1900).

Batty Langley, *New Principles of Gardening* (J. Pemberton, 1728).

Alexander Le Blond, *The Theory and Practice of Gardening* (Bernard Lintot, 1712).

Christopher Lloyd, *The Well-Tempered Garden* (Collins Publishers, 1970).

Russell Page, *The Education of a Gardener* (Collins Publishers, 1962).

Charles A. Platt, *Italian Gardens* (Harper Bros., 1893).

Julian Raxworthy, *Overgrown: Practices Between Landscape Architecture & Gardening* (MIT Press, 2018)

Humphry Repton, *Observations on the Theory and Practice of Landscape Gardening* (J. Taylor 1803).

William Robinson, *The Wild Garden* (J. Murray, 1895).

Geoffrey Scott, *The Architecture of Humanism: A Study in the History of Taste* (Constable & Co., Ltd., 1914).

H. Avray Tipping, *The Garden of Today* (Martin Hopkinson, Ltd., 1933).

H. Inigo Triggs, *Formal Gardens in England and Scotland* (Chiswick Press and Charles Wittingham & Co., 1902).

Edith Wharton and Ogden Codman, *The Decoration of Houses* (Charles Scribner's Sons, 1897).

ACKNOWLEDGMENTS

My most sincere thanks to Jeff Poole for his unwavering commitment to this book project. It has been my good fortune to have shared this experience with such a kind and genuinely talented man.

My thanks to Elizabeth White and her colleagues at Monacelli Press for embracing this book from the start. Your steady hand and formidable powers in the art of subtraction have made this the book I hoped it would be. Thanks to Susan Evans for her efforts in bringing together a challenging collection of images and drawings and designing an elegant book around them

For my clients who have allowed me the privilege of pursuing my chosen path: thank you for your kindness and generosity. I can only hope that our work together has enriched your lives as it has my own.

My deepest thanks to my longtime colleagues Patricia Hotz, Paul Gaertner, and Peter Lorenzoni. Very little of the work represented in this book would have come to fruition without your devotion and commitment to a common purpose. Thank you for traveling along this garden path with me for so many years.

I am grateful to Sam Abell for his willingness to write the foreword to this book. It has been my great fortune to know you as my teacher and my friend. Poetry.

For Charlotte Tieken whose love, support, and unbending curiosity makes this life in the garden worth living.

Virginia Donelson, *Portrait of Charles Stick.*

ILLUSTRATION CREDITS

Sam Abell 111, 118, 197
Alamy Inc. 8
Bryan Family Archives 39
Chuck Hilton Architects 72, 91, 92
Churchill Family 10
Colonial Williamsburg Foundation 17
Virginia Donelson 207
T. S. Elliott 5, 34, 35, 36, 37, 38, 40, 41, 43, 44, 45, 46, 48, 49, 51, 52, 120-121, 122, 123, 124, 126, 127, 128, 129, 138, 139, 141, 142, 143, 190-191
Roger Foley 1, 54, 55, 56, 57, 60, 61, 62, 63, 64, 65, 66, 67, 162, 163, 164, 166, 167
Gus Ford 80
Paul Gaertner 32-33, 102, 106, 111, 116
Tom Gleason 51
Dan Grogan 14
James River Garden Club 160
Jeremy Green 14
Kathy Landman 88
Larry Lederman 22 (below), 23
Chris Little 152, 153, 154, 155
Peter Lorenzoni 58, 77, 86, 95, 99, 105, 136, 156, 199
Bill Mauzy 168, 170, 172, 173, 183, 186, 187
Middlebury College 11, 12
Middlebury Special Collections 10
Prince Morrow 201
Jim Murray 182, 184, 186, 188, 189
Eric Piasecki/Otto 144, 146, 148, 149, 150, 151
Office of Charles J. Stick 59, 74, 75, 125, 133, 135, 140, 147, 154, 159, 160, 165, 171, 185, 196, 198, 199, 200, 203
Jeff Poole Cover, 21, 27, 29, 53, 69, 70, 71, 75, 76, 78, 82, 83, 84, 89, 93, 95, 96, 97, 98, 100, 104, 119, 195, 197, 203
Richard Robinson 161
Carolyne Roehm 203
Sebastian and Associates 137
Seilheimer Family 56
Charles J. Stick 2, 5, 7, 9, 10, 12, 14, 16, 18-19, 20, 22, 24, 25, 26, 28, 30, 31, 60, 61, 71, 72, 79, 85, 86, 89, 91, 93, 94, 97, 103, 105, 107, 108, 109, 110, 112, 113, 114, 115, 116, 117, 119, 130, 131, 132, 133, 134, 134, 145, 148, 150, 157, 158, 158, 161, 169, 172, 174, 175, 176, 177, 178, 179, 180, 181, 184, 196, 198, 199, 200, 202
Curtice Taylor 68, 81, 87, 94, 101
Edward Thomas 27
University of Virginia 13
Waverley Farm 106

Library of Congress Control Number: 2024906843
ISBN: 978-158093-644-6

Captions:
Page 1: Garden of the Four Seasons at Mount Sharon.
Page 2: Rose arbor at Waverley.
Page 5, from left: Harris Garden, Crab Tree Farm, Waverley.
Pages 32–33: Serpentine hedge at Waverley.
Pages 120–21: A Chippendale gate within a hornbeam hedge at the Harris Garden.
Pages 190–91: Arbor at the Winnetka Garden.

Design: Susan Evans, Design per se

Printed in China

Monacelli
A Phaidon Company
111 Broadway
New York, New York 10006

www.phaidon.com/monacelli

A WARM SUNNY DAY IN THE LOIRE – A WEEK GONE BY IN FRANCE.

I AM NOT SURE THAT WE HAVE MOVED VERY FAR FORWARD IN THREE HUNDRED YEARS.

160' PERHAPS?

LA PAGODE DE CHANTELOUPE ~ 25 MAY 1993

BLOIS ~ THE CATHEDRAL
24 MAY 1993.

25 MAY 1993
CHAMBORD